Dedicated

To

My parents

&

Wife

# Preface

This book presents a theoretical as well as practical approach to the Dynamic Web Development. It is intended to provide the readers not only with a good understanding of all the components of ASP programming, but also with a real sense of how they actually fit together to make a web application. The simple yet strong, easy yet effective, compact yet complete descriptions of every topic will surely help students to build confidence and interested towards the web programming. The book gives a large number of ready to code examples in a simplest language which helps students to digest well.

### Chapter 1:- Introduction

This chapter is an introductory chapter to the Internet. How the Internet works and How can we access the Internet are the covered topics.

### Chapter 2:- Applications of the Internet

This chapter emphasizes on the real life applications of the Internet.

### Chapter 3- Introduction to HTML

This chapter covers HTML (Hyper Text Markup Language) and CSS (Cascading Style Sheets) in detail. It discusses the static web page design.

### Chapter4- Components of Dynamic Web

This chapter is an introductory chapter to the dynamic web page design. It covers VBScript, XML(Extensible Markup Language).

### Chapter 5 – Active Server Pages

This chapter is an introductory chapter to the ASP technology. How the ASP technology works and How can we use it to develop web applications are the covered topics.

### Chapter 6- Communication with the user

This chapter emphasizes on the detailed description of ASP technology programmatically.

### Chapter 7- Database handling with ASP

This chapter shows various ways to connect with the database through ASP technology.

# Acknowledgement

It is my pleasure to take this opportunity to thank all those who helped me directly or indirectly in completion of this book. Not everything that I have received can be acknowledged with a few words but I am trying.

I am thankful to the god without whose grace and blessings nothing is possible for any one.

I extend my gratitude to my parents and wife who encouraged me with their blessings throughout my life. I also like to thank all the faculty members of Government Engineering College Rajkot who appreciated the way I work.

Finally how can I forget my beloved students? I want to thank all of them for liking and appreciating my teaching.

The readers of the book are encouraged to send their comments, suggestions at *hardik.molia@gmail.com*

Prof. Hardik K. Molia

# About The Author

Prof. Hardik K. Molia is a Computer Engineer by Education & an Assistant Professor by Profession...

He is working at Government Engineering College, Rajkot, Gujarat, India...

His specializations are Operating Systems, Compilers, Theory of Computation, System Programming, Network Security, Parallel Processing...

You may contact him, hardik.molia@gmail.com

# Index

# Chapter 1:- Introduction

## 1.1 INTRODUCTION TO INTERNET

Since one decade, General purpose computers and other computer related devices are available at very affordable cost and because of their other advantages; it is directly or indirectly becoming an integral part of almost every human activity. It is natural that people want to connect computers to share information. By connecting computers together, not only it is possible to share information, but also a user at some location can access and use computers that are physically located quite far away.

### 1.1.1 Computer Network:-

A computer network is an interconnected collection of independent computing resources. When we say that two computers are interconnected it means that both computers should be able to communicate with each other. Networks are broadly classified according to the geographical area that they span. By this criterion, network can be classified as a

Local Area Network (LAN)

Metropolitan Area Network (MAN)

Wide Area Network (WAN)

LAN connects many computers in a limited geographical area (single office, a building or a campus).

MAN connects many computers in a network which spans over a city.

WAN connects many computers (many MANs or LANs) and extends the network over countries.

Internetworks or internet (note the lowercase letter I – internet and Internet both are different).

When two or more networks are connected with each other, they become an internetwork or internet. It is possible to connect two or more networks of based on different technologies.

### 1.1.2 Internet:-

Imagine, we are in a room full of people from different countries and everyone knows his native language only. In order to communicate, we have to set some standard rules and vocabulary that everyone can speak and understand. Similarly the Internet is a system that sets some common rules to make communication possible among different computer networks.

The Internet is collaboration of thousands of interconnected networks. The Internet is a global collection of computers and other devices, which are linked through computer cables and telephone lines, making communication possible with each other to share computer equipment, programs, messages and information. In other words, it can be defined as networks of networks as it connects many incompatible (developed using different technologies) LANs and WANs.

Internet allows you to share resources. One such major resource is information, which exists in computers in the form of files or database. Thus, one of the key aspects in networks of many computers is to move the files between two specific computers.

Private individuals as well as various organizations (such as government offices, universities, colleges, and companies) in more than 100 countries use the Internet.

No one owns the Internet, Companies and individuals may own a small part of the Internet, such as servers and the connections between them but there is no one that owns it all. There are some organizations that maintain and standardize what happens on the Internet such as such as the National Science Foundation, the Internet Engineering Task Force, ICANN, InterNIC and the Internet Architecture Board.

### 1.1.3 Intranet:-

There are also small networks known as Intranet. It is mainly use by organizations or by group of people. Using Intranet branch offices of organization can connect to their head office for sharing of data. This makes centralized data collecting server. Data can be collected from any of the branch office.

### 1.2 HISTORY OF INTERNET:-

The origins of the Internet lie in the cold war. A network was implemented to connect military computers in United States to transfer important messages during the war. These computers were linked together using telephone lines. In those days, the network was based on the circuit switching networking.

In 1960, computers from different manufactures were unable to communicate with each other. So the Advanced Research Projects Agency (ARPA) started work behind this problem.

In 1969, ARPA introduced ARPANET which connected US universities. ARPANET was the first network based on packet switching networking. This grew into the UK academic network, JANET.

In 1972, ARPA introduced TCP (Transmission Control Protocol) to make the communication possible between two incompatible networks through a device named as gateway.

In 1973, ARPA spited TCP in to two parts. TCP and IP and that was the development of TCP/IP internetworking protocol.

In the 1980s the various networks began to be linked together, so we could get from one network to another. This is when the name "Internet" was first used.

The Internet today is not a simple hierarchical structure. It connects many WANs, MANs and LANs (of different technologies) using various connecting devices and switching stations. It is very difficult to give exact representation of the Internet because it is continuously changing

## 1.3 INTERNET SERVICE PROVIDER

Internet Service Provider refers to a company that provides Internet services for personal use to business use. The service provider provides a software package, username, password and access phone number. Equipped with a modem, you can then log on to the Internet and browse the Internet.

The Internet access is managed by a hierarchy of ISPs.

**International Service Providers:-**

These ISPs connect various nations together.

**National Service Providers:-**

NSPs are networks created and managed by specialized companies to provide connectivity between the end users. These networks are connected by complex switching stations called network access points. NSPs operate at higher data rate.

**Regional Internet Service Provider:-**

Regional ISPs are smaller ISPs that are connected to one or more NSPs. They have lower data rate as compared to NSPs.

**Local Internet Service Provider:-**

Local ISPs provide direct service to the end users.mostly end users are directly connected with the local ISPs. Local ISPs are connected to regional ISPs or directly to NSPs.

## 1.4 DOMAIN NAME SYSTEM:-

To reach to a person, we go to his address. Similarly to identify each device uniquely or to communicate with it on the Internet, it uses the numerical representation of an address known as IP address. But it is difficult to remember numerical values and so people prefer to use alphanumeric names. Domain Name System maps an alphanumeric user friendly name to corresponding IP address and vice versa.

It is required to store this kind of information. it is very inefficient to store all the names and their corresponding IP addresses on one computer because responding to requests from all over the Internet becomes a heavy load on the server. The solution of this problem is to distribute the information among several computers called DNS server.

The names must be unique because addresses are unique. Each name is made of several parts. (1st part specifies nature of the organization, 2nd part defines the name itself, 3rd part defines branch name and so on.).

All parts are arranged in a hierarchical tree representation. A name is the concatenation of all the parts from root to children separated by a ".".(dot)

A domain name space is the set of rules that we need to follow while deciding a domain name for an entity.  The tree representation of the name is the domain name space.

On the Internet, we use DNS to identify a website on a web server. The domain name space is divided into three different sections. An identification label, Generic domains and country domains.

- Identification label:-
    The identification label identifies the entity by a name. It can be the name of the university, name of a company or name of an individual.

- Generic Domain:-
    The generic domain uses two or more characters for organizational abbreviation. It defines registered entities according to their behavior. According to the type of services and content provided by that entity.  Some of the popular labels for generic domains are listed below.

| Generic domain label | Description |
|---|---|
| com | Commercial organization |
| edu | Educational institutes |
| gov | Government organization |
| org | Non profit organization. |

- Country domains:-

The country domain uses two characters for country abbreviation. It indicates the country to which the entity belongs). It can be further classified by specifying by using two characters for national designation. For India, we can use "in", for United States we can use "us".

Example:- india.gov.in    is the website.

<u>india</u>   is the identification label which indicates this web site is  related to India.

<u>gov</u>   is the generic domain which indicates the site is owned by a government organization.

<u>in</u>    is the country domain which indicates the site is from the country India.

## 1.5 URL:-

The URL is the address of a particular file on the Internet. This address can be part of either a domain name or an IP (Internet Protocol) address. The URL is the logical path on the server. We will discuss more about the logical paths in chapter 5.  A URL is made up of three or four components as shown below.

http://www.xyz.com/data/pages/index.asp?t1=hardik+molia&p1=qsxrfvyhn&r1=male

## A scheme:-

The scheme specifies the protocol which is used to access the page on the internet. Here it is http(hyper text transfer protocol)

## A Domain name or ip address:-

The next component is either the domain name of the server or the actual ip address of the server. If a server is hosting more than one web applications, port addresses are used to distinguish among all applications. Here it is www.xyz.com

## Path:-

The path is the virtual path which identifies a specific file on the server which client wants to access. Here it is data/pages/index.asp

## Query String:-

The query string is a string of name and value pairs which the web page uses for special purpose. It is one of the methods of passing information to the page before it starts execution.

Here it is t1=hardik+molia&p1=qsxrfvyhn& r1=male

'?' indicates the beginning of the querystring.

'&' separates each component of the querystring.

'+' indicates a ' ' (space) in querystring.

t1=hardik+molia indicates the value of t1 field is "hardik molia"p1=qsxrfvyhn indicates the value of p1 field is "qsxrfvyhn" and so on...

During programming, we often need to specify the URL of various files stored on the server.(other web pages, images, sound files etc.) there two ways we can specify a URL – absolute URL and relative URL. Consider the following directory structure.

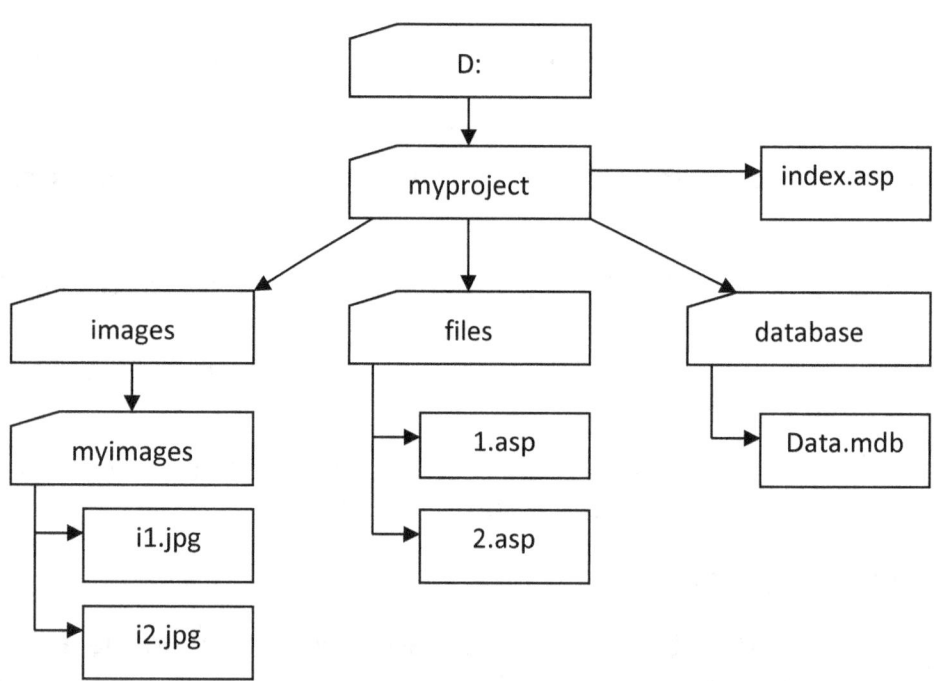

D: drive has one folder "myproject" which has three folders (images, files and database) and index.asp file. Files in different folders are shown in above figure.

**Absolute URL:-**

An absolute URL is the path which we write with reference to the root directory

D:\myproject\images\myimages\i1.jpg        valid URL as per the above directory structure

D:\myproject\files\index.asp        valid

D:\myproject\database\2.asp        invalid

Absolute URLs are very easy to use but because we write such URLs with reference to the root directory, if we change the entire directory "myproject" E: or to some other location, all the absolute URLs will become invalid because they still refer to D:

**Relative URL:-**

A relative URL is the path which we write with reference to the presently working(current) directory. In relative URL a "." (dot) is used to represent the current directory. And ".." (dot dot) is used to represent one directory up from the current directory.

The table below shows some combinations of absolute and corresponding relative URLs. Assume that we are presently working in D:\myproject\files

| Absolute URL | Relative URL |
|---|---|
| D:\myproject\files\1.asp | 1.asp |
| D:\myproject\files\1.asp | .\1.asp |
| D:\myproject\index.asp | ..\index.asp |
| D:\myproject\images\myimages\i1.jpg | ..\images\myimages\i1.jpg |

We want to represent D:\myproject\index.asp as a relative URL from D:\myproject\files

The index.asp file is stored inside the folder "myproject" which is one directory up from the files directory. So we use ".." to move to one directory up("myproject") from the current directory ("files") and \index.asp specifies the index.asp file stored inside the "myproject" directory. So the relative URL is ..\index.asp

We can use ".." more than once if we want to move to more than one level of directories up.

**1.6 WEB SERVER**

A web server is a combination of computer hardware and software which delivers web pages to client. Every web server has its unique IP address and a domain name. The web browser requests to the server using the URL (URL contains information about either the IP address or the domain name) to the server. Server finds the requested page and sends it to the browser.

Web servers are responsible for storing and exchanging information with other computers so at least two computers are required for exchange of information, a client who requests the information and a server which provides the information. Each side requires software to negotiate the exchange of information.

Client requires a web browser like internet explorer, mozilla, opera etc.

15

Web server requires software to negotiate the data transfer between clients and server via various communication protocols. The type of server software used by a web server depends on the operating system of the web server. For example, for windows, Internet Information Services(IIS) is a popular choice while for UNIX systems, apache web server is popular.

A simple exchange between the client machine and Web server goes like this:

The client's browser sends its request to the server using a URL which includes protocol, domain name, path and page name.

A domain name server translates the domain name into its corresponding IP address and forwards the requests to the server.

The server sends the request to the web server for further processing. The web server verifies that the given page(or address) exists, finds the necessary files, runs the scripts and returns the results in HTML form, back to the browser. If the server can not locate the file, the server sends an error message to the client.

The client browser displays the HTML content received from the server.

This process is repeated until the client browser leaves the site.

## 1.7 HTTP

HTTP stands for Hyper Text Transfer Protocol. It is one of the Application layer protocols of OSI network model.

HTTP defines how client and server transfer documents. it also defines actions taken by client and server for various commands.

In client-server computing, HTTP works as a request-response protocol. A user's web browser works as a client. Client requests for a web page by sending a HTTP request to the server. The server finds the web page, executes it(if required) and sends the response back to the client.

An HTTP session is a sequence of request-response communications. A client initiates the communication by sending a HTTP request. It establishes a Transmission Control Protocol (TCP) connection to the server. An HTTP server receives client's request and sends back a status line, requested response or an error message.

HTTP is a stateless protocol because each request-response is executed separately so it is not possible to maintain the state between two requests through HTTP only.

HTTP defines various methods (GET, POST, TRACE, PATCH, HEAD etc.) to collect information from the user and to send the information to the server.

HTTP's secure version is HTTPS, which is used to delivery the sensitive content to the server in encrypted form.

## 1.8 REVIEW QUESTIONS:-

1. Write short notes on following.

   - The Internet
   - Internet Service Providers
   - Domain Name System
   - URL
   - Web Server
   - HTTP

2. Explain the difference between absolute URL and relative URL with examples.
3. Explain how the web server works.

# Chapter 2:- Applications of The Internet

## 2.1 WORLD WIDE WEB

People use the terms "Internet" and "World Wide Web" or "Web" in every day life without much distinction. However both are not same. The internet is a large network of interconnected computer networks while the web is an application running on the Internet. Web is a collection of interrelated and interlinked documents and other resources identified by URLs and interlinked by hyperlinks.

A user enters a URL of the page in the address bar of the web browser to access from the web. The browser then requests the page by sending a HTTP request to the web server.

The www is the information sharing model which uses HTTP protocol to make communication between a client and a web server.

The Internet itself is a network. There is no use of a network, if we don't have any data to share among the network. Te Web provides a way of storing and accessing information scattered across millions of computers connected to the Internet around the world.

Before the web, it was very difficult (using UNIX commands or specialized tools) on the Internet to access a file. Web introduced a very user friendly way to access the Internet.

The www is widely spread across the world and it is very complex(millions of computers are part of the www) just like a web and so it was named as World Wide Web.

## 2.2 WEB SITES

Individual documents which we write are called Web pages, and a collection of related documents, arranged in a specific sequence is called a Web site. People visit pages of web sites to get various kind of information. Today we can find almost everything on the Internet. Here is a list of services available on web sites.

Personal Information, Educational, Business, Shopping, Entertainment, Social Networking, Emails and instant messaging ,Search Engines, Government Services, File Hosting Sites

There are many programming languages are available to write web pages. After writing web pages, we do interlinking of web pages to make it possible to access other pages from a page.

Each page has its own unique URL. The client requests for a page either by specifying the URL of the page or by following any link to the page.

A web site is a interrelated collection of web pages with some other files like images, video, database files etc.

Basically there are two types of web sites we can develop. Static web sites and dynamic web sites.

| STATIC WEB SITES | DYNAMIC WEB SITES |
|---|---|
| Programmer writes web pages and the contents of these pages are fixed for all clients. | Programmer writes web pages and<br><br>But contents of these pages are not fixed. The webserver deicides the contents according to the client requests. |
| We use HTML for static web page design | We use various technologies like ASP, ASP .Net, PHP etc for dynamic web page design. |
| A client requests for a page with URL of a page only. | A client requests for a page with a URL of a page and some other information. Based on this information, the web server will put the content in the page. |
| A web site which provides online reading facility of articles is the example of static web site. A client can only view the pages developed by the programmer. | An online result system, railway reservation, Shopping are the examples of dynamic web sites. |
| Static web sites are only for the delivery of the content which is common for all the clients. | Dynamic web sites are for the delivery of the content specifically dedicated to a client and based on his requirements. |

## 2.3 SEARCH ENGINES

A web search engine is a program which we can use to search specific information on the World Wide Web. The information can be web pages, images, audio or any other files.

We can search by specifying the information to search by entering our criteria in text. The search engine searches it and provides a list of matched information.

The result of the search is a collection of URLs(web pages on WWW) which contains the required information.A search engine works in three phases. Web crawling, Indexing and Searching.

A Web crawler is a program that browses the pages of the World Wide Web automatically in an ordered manner.

Search engine indexing collects, analyzes and stores data of all the pages visited by Web crawler. It maintains a database of search engine which contains data about what information is available on different web pages of different web sites.

A web search query is a sequence of words that a user enters into web search engine to search. A user can search for a topic, web page, web site or any other resources.

The search engine searches the data from its database to get the list of URLs which contains the required information.

Crawler-Based Search Engines are very effective because they maintain the latest database automatically so mostly at any time it is updated with the latest information on the World Wide Web.

Human-Powered Directories are also search engines but they don't maintain the latest information of web pages automatically. They provide the search facility from the data given by various web sites. So it is comparatively less intelligent.

It is possible to get the different search result for the same search on different search engines because the database maintained by a search engine and the program behind a search engine are different from other search engines.

A Meta search engine sends the user requests to other search engines and provides the consolidate result which it composes by merging results provided other search engines too.

The search is the time consuming process, to delivery the search result as fast as possible, the peer to peer search engines are used. These engines are based on the distrusted application architecture in which a single task is partitioned among more than one computer.

Google, yahoo, ask.com, bing are the most popular search engines.

## 2.4 ELECTRONIC MAIL

The electronic mail (e-mail) is one of the methods of exchanging information from one user (sender) to one or more users (recipients).

In earlier days of the Internet, email systems were required that sender and recipients must be online at the time of communication.  But nowadays, email systems are implemented on a store and forward model. Email servers receive the emails, store them, forward them and deliver them.

The sender needs to remain online as long as the browser takes time to send an email. The recipients need to remain online as long as the browser takes time to receive an email.

An email box is similar to the letter box of post offices. It is the place on the server where emails are stored electronically.

An email address identifies an email box for a user. Each user (either sender or recipient) has his own email address provided by email service provider while creating an email account.

For security, an email account is accessible only using a valid username and a password.

The format of an email address is abc@xyz.com where "abc" can be anything selected by the user(no spaces are allowed) which identifies an email box. The string after the "@" specifies the location of the email box. xyz.com(to the server whose domain name is xyz.com)

Simple Mail Transfer Protocol (SMTP) protocol is used to send only text based emails to the email box, Multipurpose Internet Mail Extensions (MIME) protocol extends the format of an email by allowing non-text content(binary files as attachments, images, audio etc.)

Mailboxes are access by using either the Post Office Protocol (POP) which provides one way communication from mailbox to the recipient or using the Internet Message Access Protocol (IMAP) which provides two way communication between mailbox and the recipient.

POP downloads the entire email to the recipient's machine while it is possible to download only header of an email to the recipient's machine using IMAP.

Using IMAP, we can also mark emails as read or unread and email box modifies the status accordingly.

SMTP, MIME, POP and IMAP work on the Application layer of the OSI network model.

An email consists of three components, the envelope, the header and the body.

The header of the message contains sender's email address, email addresses or one or more recipients, subject of the email, date/time on which the email is being sent.

## How email works?

Suppose user Anil's email address is anil123@xyz.com Anil wants to send an email to Ravi whose email address is ravi456@abc.com

Anil composes an email address using the editor provided by xyz.com  or using any other email client application like outlook (known as Mail User Agent -MUA). He enters ravi456@abc.com in the field of recipient's email address and presses the Send button.

The MUA running on anil's computer forwards the email message using SMTP or MIME protocol (based on the content of the email) to the Mail Submission Agent (MSA). MSA is a computer program mostly running on ISPs which receives emails from the clients and processes them.

The MSA receives the email message. It identifies the recipient's address which is ravi456@abc.com using DNS revolver it finds the address of the email server corresponding to the DNS abc.com and forwards the email message to that address.

The email server receives the message. It stores the email message into the email box which is identified as ravi456. The email is now accessible by ravi.

Ravi can open his email client (MUA) by entering a valid user name and password. He can access his email box and can download Anil's email.

## 2.5 WEB PORTAL

A web portal is a web site which provides many ways to access the Internet from one site. A web portal provides access to the WWW, discussion forums, social networking, search engine, email services, news, entertainment, stock prices etc. AOL, MSN, yahoo!, iGoogle are popular web portals.

A horizontal web portal covers many areas in it. It can be used as a common platform for several companies. A vertical web portal is a specialized entity dedicated to a specific area.(a company, a subject, an interest etc.)

A personal portal is created and managed mostly by individual person. It is used to represent personal information on the server. Mostly personal portals contain services like email services, discussions, chats, social networking etc. a personal portal can also refer to a company or an institute.

A business portal is designed to share collaboration in workplaces. A business portal should be easily and properly accessible from multiple platforms such as personal computers, cell phones, PDAs etc.

A news portal is dedicated to provide latest news to its visitors. A more advance ways of news portals like searching a news, news with videos, and news with other references are very popular among the people. People can also submit any news or they can participate in online polls.

A government web portal is the place from where citizens can get all the information that they get normally by visiting a government sector. The portal shows the process of acquiring various government documents (passport, driving licese etc.) people can request for government services or can get the current status of their requested services.

A Stock portal shares all the stock market related information to all the share holders. People can find latest price, ask/bids, reports and latest news regarding shares. Some stock portals also provide services to buy or sell shares through the Internet.

We can say that a web portal represents a specific community on the Internet. We can broadly define the categories of web portals based on the communities they refer, as shown below.

- **Consumer-focused portals:-**

    Geographic Portals
        Based on any specific physical location of the world.
        Ex:- a portal for Indian people.


    Demographic Portals
        Based on gender, caste, language etc.
        Ex: a portal for teenager, a portal for senior citizens etc.

Topical Portals
Based on topics of interest
Ex: a portal for music, paintings, dancing etc.

- **Business-to-business portals:-**

Vertical industry portals
Focused on an industry segment
Ex: a portal for softwares, automobile, consumer electronics etc.

Functional portals
Focused on a business
Ex: a portal related to marketing, purchasing, selling etc.

## 2.6 CHAT

The technical term of chatting is the Instant messaging (IM) it is the real time text based communication between two or more users (working on different computers or on devices like mobiles) on the Internet.

Users can be a part of a chat through IM client programs(yahoo messengers, google talk etc.) a user needs to enter a username and valid password at the time of login to the IM client program.

The chat is real time. It means all the users who want to participate in a chat, must be online at a time and using a IM client software.

Initially the instant messaging was purely text based. Nowdays we can also do voice chanting and video chatting which sends and receives voice and video respectively.

IM provides an effective, efficient and immediate way of transferring messages to other users; IM allows instant reply from the recipients. IM client provides various facilities like,

- ✓ Managing groups of friends
- ✓ Notifications of online friends (those who are currently using the same IM software)
- ✓ Making ourselves invisible on the IM.
- ✓ Communication using web cam or head phone.
- ✓ Chat Rooms based on gender, interest, education, age etc.

## 2.7 TELNET

Telnet is an application layer protocol which provides a service to access remote computers through the Internet.

Using Telnet, a user can access someone else's computer from the remote place (from his own computer). He needs to execute telnet command from his computer with the DSN or IP address of the remote computer.

telnet 192.178.32.2        or        telnet www.xyz.com

The command prompt asks for the authentication. If user has valid username and password, he can access the remote computer successfully.

Using HTTP or FTP protocols, we can only request for specific files from remote computers but can not be logged on remote computer just like one of the users of that computer can. Using TELNET, we can log on to the remote computer just like any regular user of that computer can.

Using HTTP or FTP, we don't have any privileges to modify or delete the files, using TELNET we can modify or delete files also. We can also execute other programs on remote computer from our computer.

## 2.8 FTP SERVERS

File Transfer Protocol (FTP) is an application layer protocol which is used to transfer file from one computer to another computer over a TCP/IP based network. The Internet is a TCP/IP network and so FTP is widely used on the Internet. The FTP server is the server which supports file transfer services using FTP protocol.

Server can use FTP protocol to transfer files to clients in two ways.

Server asks for the authentication (username and password) and on identifying an authentic user, server allows file transfer using FTP.

Server does no ask for the authentication. Files are accessible to all the clients through FTP.

In earlier days of the Internet, text based commands were used to transfer files. But nowadays GUI based applications are available which transfer files using FTP.

## FTP modes:-

FTP can be used in either Active mode or in passive mode which specifies how the data connection is established, maintained and terminated.

In Active mode, client sends its address (ip address and port address on which client receives files) to the server. Server establishes the TCP connection to that address.

The passive mode is used if client is behind the firewall and unable to accept files from incoming TCP connections. At this time, client sends a PASV command to the server. Server sends its own address (ip address and port address from which server sends files) to the client. Client establishes the TCP connection to that address.

## Modes of Data Representation:-

ASCII Mode:- this mode is used only or files containing plain text. If the actual data is not in ASCII form, server converts it to the ASCII form and sends it. The client receives the data in ASCII form and converts back to its original form.

Image Mode (Binary Mode):- the server sends the file byte by byte.

EBCDIC Mode:- this mode is similar to the ASCII mode but uses EBCDIC code to represent text based content.

Local Mode:- this mode requires an identical setup of a specific format on client and server side. The predefined format is used to transfer files.

## Modes of Data transfer:-

Stream mode:-This mode transfers data as a continous stram of bytes. FTP is only used to initiate the transfer then all processing is left up to TCP.

Block mode:-FTP breaks the data into several blocks and transfers the data block by block using TCP.

Compressed mode:- before sending the data, FTP server compresses the data and then sends either using stream mode or using block mode.

FTPS (Secure FTP) is a protocol to transfer files with security. FTPS is used to transfer the encrypted version of files for the security purpose.

## 2.9 PROXY SERVERS

A proxy server is a computer program which is an intermediate entity between the clients and the servers.

The proxy server receives all requests of a client such as a web page, a file or other resources. The proxy server has some database with keeps filtering rules.

The proxy server decides whether to forward the request of the client to the server or not based on the filtering rules. If the proxy server finds a valid request, it forwards the request to the relevant server on behalf of the actual client and provides the response of the server back to the actual client.

A proxy server can alter the client' request and sometimes it can send the response to the client without contacting the server. The proxy server can cache (store) some commonly used responses and can directly use them sometimes.

A proxy server which requests and replies without any modification is called a gate way or tunneling proxy server.

A proxy server has a large number of purposes as listed below.

- To implement security while accessing other networks or the Internet
- To speed up the access using caching of commonly used resources (files or web pages)
- To block undesired web sites.
- To track Internet usage of clients.
- To scan the incoming content for viruses before transferring to the clients.
- To scan the outgoing content for illegal data leakage.
- To implement parental controls or regional restrictions.

### Types of proxy servers:-

Forward proxy server:-

A Forward proxy server works inside the local network to which the client belongs. The forward proxy server is accessible by only the clients of that network. The clients belong to that network always need to access the same forward proxy server.

Reverse proxy server:-

A Reverse proxy server works inside the local network to which the server belongs. The clients access the proxy server only when they want to access the web server corresponding to that proxy server.

28

Open proxy server:-

An open proxy server is a separate entity on the network which does not belong to any of the network of clients or of the server. The clients and the server access the proxy as it is an independent server.

## 2.10 NEWS SERVERS

A news server is used to handle Usenet articles. The Usenet is a discussion based network on the Internet. Using Usenet, Users read and post messages known as articles, post or news on the Internet.

Usenet organizes the articles into categories. Each category is known as a newsgroup. Each newsgroup can be organized into hierarchies of subjects. For example software.asp and software.php are under the software hierarchy, each category can be further divided into sub categories if required.

A news client or a newsreader is an application program, used by the clients to access the Usenet. This program reads or posts user articles and keeps track about user's activities. A news client uses Network News Transfer Protocol (NNTP) to communicate with the Usenet. NewsBin, NewsLeecher, Xnews are popular news client applications.

The articles are stored on various news servers. These servers remain in communication continuously to exchange the latest articles. A news server can work as a reader server or a transit server or both.

A reader server provides facility to read or post articles from news client software. A transit server exchanges articles with other news servers.

When a user posts a new article or edits an existing article, initially it becomes available on the user's news server. Each news server communicates with one or more other news servers and exchanges articles. So after some time, the article will be copied from one news server to other news server and sometimes to all news server. This way the articles of all users are maintained on the Internet.

The format of articles is similar to the format of emails. But both are not same. We send an email to one or more specific recipients only. The article is the broadcasting of a message to the world or to the group.

## 2.11 DIRECTORY SERVERS

A directory is the storage of some data. A directory service is software which stores information about network's users and resources. A directory server is the place where we store directories.

A Directory Service Agent (DSA) is a program which manages access to the directories for users. There are various directory services are available based on their purposes.

A naming service, maps the name of network resources to their network addresses. While using a naming service, user don't need to provide physical or logical addresses, by providing a resource name, user can access it.

The Lightweight Directory Access Protocol (LDAP) is the application layer protocol used to access directory services over the Internet.

Using LDAP, clients can add, modify, delete entries. Clients can search, compare and filtering of entities.

## 2.12 REVIEW QUESTIONS:-

1. Write short notes on following.

   - The World Wide Web
   - Websites
   - Search Engines
   - Electronic Mail
   - Web Portals
   - Instant Messaging (Chat)
   - Telnet
   - FTP servers
   - Proxy servers
   - News servers
   - Directory servers

2. Explain the difference between static web sites and dynamic web sites.
3. Explain the applications of the Internet.

# Chapter 3:- Introduction to Html

## 3.1 WHAT IS HTML?

- HTML stands for Hyper Text Markup Language.
- HTML is not a programming language, it is a markup language.
- Hypertext is the text displayed on a computer with references (hyperlinks) to other text that the reader can access usually by a mouse click or key press events.
- A markup language is a combination of words and symbols which give instructions on how a document should appear on web browser.
- HTML is a small subset of full featured Standard Generalized Markup Language(SGML)
- HTML contains two kind of information:-
  (1)Markup:-  All the text contained between angle brackets <>
  (2)Content:- All the text not contained between angle brackets <>
  Markup information tells the browser how to display the content.

- HTML is not a programming language so it has no programming language features like decision making capability.
- HTML is a case insensitive language.
- HTML treats more than one space as a single space and ignores line break.

## 3.2 SYNTAX OF HTML DOCUMENT:-

A HTML document is the representation of some content using markup. To represent markup, HTML has various tags.

### Tags:-

Tags are tags of the HTML document used to specify how the document should be displayed by the browser. In HTML, each tag has its own specific meaning. Examples:-tags for text formatting, images, tables, frames etc.

### Attributes:-

Attributes are associated with each tag to further define the tags. Attributes are the characteristics of tags. Examples:-size, color, background, bgcolor etc.

The general syntax of a tag is as follows:

<tag attribute1 = "value" attribute2 = "value" ... > content </tag>

HTML tag(not required for all tags) has two parts, start tag and end tag.

Start tag:- <tag>indicates the beginning of the tag.

End tag:- </tag>indicates the ending of the tag.

## 3.3 EXECUTION OF A HTML FILE:-

We do not need any specific software to run HTML files on our local computers. There are many editors like Microsoft front page, macromedia dream weaver etc. are available which provide a GUI based environment through which we can design and run HTML files easily. Follow the following step to run a HTML file without installing such web editing tools.

(1) Run any text editior. E.g. Notepad
(2) Type your HTML programs
(3) Save the program in appropriate location with a valid filename and .html or .htm extension. Don't forget to select "all files" in the save as type field.
(4) We can run the HTML files on web browsers. The default web browser supported by Microsoft OS is the Internet Explorer. It is also possible to install other web browsers like Opera, Netscape Navigator, Google Chrome etc.
(5) Run a web browser.
(6) In Internet Explorer, Go to File menu of web browser -> Open ->Browse-> select your HTML file -> Press ok.
(7) You will get a web browser window containing the output of your HTML file. As shown in next section.
(8) Instead of going to steps 5 and 6.It is also possible to run the HTML file by double click on it from the folder where you have saved it.

## 3.4 STRUCTURE OF HTML DOCUMENT:-

```
<!DOCTYPE html PUBLIC "-//W3C//DTD HTML 4.01//EN"
"http://www.w3.org/TR/html4/strict.dtd">
<HTML>
<HEAD>
<TITLE>PAGE TITLE</TITLE>
</HEAD>
<BODY>
HELLO WORLD...
</BODY>
</HTML>
```

**The Doctype:-**This provides the web browser with information about the type of markup language in which the page is written. Mostly web editors create this tag automatically after the selection of document type.

**The <HTML> Tag:-**The <HTML> tag defines the beginning and the end of an HTML document. The <HTML> tag is the root tag of the document, The <HTML> tag directly contains the <HEAD> tag, the <BODY> tag, and potentially the <FRAMESET> tag instead of the <BODY> tag.

**The <HEAD> Tag: -** The head section is used to specify some additional information about the page. We can specify the text which should appear in the title bar when we open a web page. If we want to use style sheets (topic 3.13) in a web page then we write it inside this tag. It is also possible to write client side scripts here.

**The <TITLE> Tag:-**The title tag contains the text which browsers display in a title bar at the top of the browser window. This section is optional.

**The <BODY> Tag:-**Body tag contains all the content of a page. Everything that you can see in the browser window  is contained inside this tag, including paragraphs, lists, links, images, tables, and more.

3.1.html (Basic Structure of a HTML document)

| | |
|---|---|
| <HTML><br><br><HEAD><br><br><TITLE>HTML DEMO</TITLE><br><br></HEAD><br><br><BODY><br><br>HELLO WORLD...<br><br></BODY><br><br></HTML> | OUTPUT:-<br>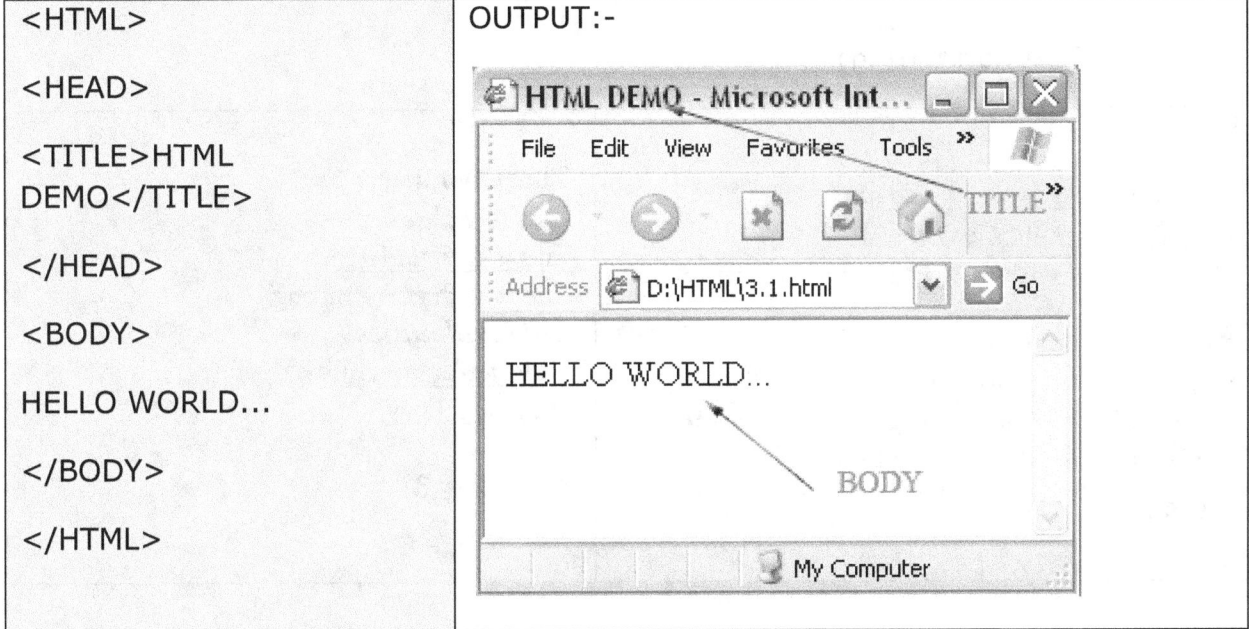 |

## 3.5 TEXT FORMATTING:-

TEXT refers to the sequence of characters (alphabets, numbers, and special symbols) which we display on a web page. We specify what we want to display on the page in the BODY area of the page. TEXT formatting tags are used to display the text more attractively by changing its font, color, style, alignment etc.

**<B>,<STRONG> :-**
These tags are used to display the text with bold font effect.

**<I>,<EM> :-**
These tags are used to display the text in italics.

**<U>:-**

33

This tag is used to display the text with underline.

## <S>,<STRIKE>:-
These tags are used to display the text with strike-through effect.

## <H1> TO <H6>:-
The H1 tag is used to specify level 1 heading. There are 6 levels of headings (H1 - H6) with H1 is the heading with the biggest font size and H6 is with the smallest font size.

| Attribute | Value | Description |
| --- | --- | --- |
| Align | Left, Center, Right | Text alignment in paragraph |

## <BR>:-
This tag is used to  break the line in web page.

3.2.html (Text Formatting)

| | |
| --- | --- |
| ```<br>\<HTML><br>\<BODY><br>\<B>This is bold\</B>\<BR><br>\<I>This is italics\</I>\<BR><br>\<U>This is with underline\</U>\<BR><br>\<S>This is with strike<br>through\</S>\<BR><br>\<B>\<I>Bold and italics\</I>\</B>\<BR><br>\<B>\<U>Bold and underline\</B>\<BR><br>Underline Only\</U>\<BR><br>\<H3>Heading 3\</H3><br>\</BODY><br>\</HTML><br>``` | **OUTPUT:-**<br><br>**This is bold**<br>*This is italics*<br><u>This is with underline</u><br>~~This is with strike through~~<br>***Bold and italics***<br>**<u>Bold and underline</u>**<br><u>Underline Only</u><br><br>**Heading 3** |

## <PRE>:-
HTML treats more than one space as a single space and ignores line break. This tag is used to d display the text in preformatted form. Browsers normally render pre text in a fixed-width font(usually courier), with white space, new lines in tact, and without word wrap.

## <TT>:-
This tag is used to display text with mono spaced effect similar to teletype text

## 3.3.html (Text Formatting)

| | |
|---|---|
| ```<HTML><br><BODY><br>HTML stands for hypertext markup language...<br><BR><br><PRE><br>HTML stands for hypertext markup language...<br></PRE><br><TT><br>HTML is not a programming language.<br></TT></BODY></HTML>``` | **OUTPUT:-**<br><br><br>HTML stands for hypertext markup language...<br><br>HTML        stands for hypertext        markup language...<br><br>HTML is not a programming language. |

**<SUB>:-**
This tag is used to define subscript text.

**<SUP>:-**
This tag is used to define superscript text.

**<FONT>:-**
This tag is used to set font, size and color of the text.

| Attribute | Value | Description |
|---|---|---|
| Color | Red, Green,Blue<br>#34D523(#RRGGBB) | Font color |
| Size | A Number From 1 To 7. Browser Default is 3. 1 Is The Smallest Size | Font size |
| Face | Verdana, Arial | Font family name |

**Example:- 3.4**

3.4.html (Text Formatting)

| | |
|---|---|
| ```<HTML><br><BODY><br>2<SUP>2</SUP>=4<BR><BR><br>H<SUB>2</SUB>O is the water<BR><BR><br><FONT FACE=VERDANA SIZE=3 COLOR=RED><br>HELLO WORLD...<BR><br>H<SUB>2</SUB>O is the water<BR><br></FONT></TT></BODY></HTML>``` | **OUTPUT:-**<br><br>$2^2=4$<br><br>$H_2O$ is the water<br><br>HELLO WORLD...<br>$H_2O$ is the water |

### <SMALL>:-
The range of the size of font in HTML is from 1 to 7. 1 is for the smallest size and 7 is for the largest size. With every <SMALL> tag, Browser will display the enclosed content with one size smaller than the current font size.

### <BIG>:-
With every <BIG> tag, Browser will display the enclosed content with one size larger than the current font size.

### <P>:-
The P tag is used to indicate paragraphs.

| Attribute | Value | Description |
|---|---|---|
| Align | Left, Center, Right | Text alignment in paragraph |

3.5.html (Text Formatting)

```
<HTML>
<BODY>
<FONT SIZE=3>
H<BIG>E<BIG>L<BIG>L<BIG>O<BR>
w<SMALL>O<SMALL>R<SMALL>L
<SMALL>D<BR><BR>
<P ALIGN=CENTER>
This is the world of computers in which
everything is possible...</P>
</BODY>
</HTML>
```

OUTPUT:-

ʜᴇLL O

WORLD

This is the world of computers in which everything is possible...

## 3.6 LISTS:-

In HTML, it is very easy to display a set of items in a list form. HTML lists are of two types: Ordered list and unordered list. An ordered list is a list with numbers or alphabets assigned with each list item. An unordered list is a list with some symbols(circles, squares) assigned with each list item.

### <LI>:- (LIST ITEM)
This tag is used for a single list item. We use this tag inside either <OL> or <UL>.

### <UL>:- (UNORDERED LIST)
The UL tag creates an unordered list. The tag contains one or more LI tags that define the actual items of the list.

| Attribute | Value | Description |
|---|---|---|
| Type | Disc,Square,Circle | Specifies the style of the bullets of list items. |

3.6.html (Unordered List)

| | |
|---|---|
| ```<br><HTML><br><BODY><br><H3>D.I.T.-V</H3><br><UL TYPE=CIRCLE><br><LI>WP with VB .Net</LI><br><LI>Wireless Communication</LI><br><LI>Dynamic Web<br>Development</LI><br><LI>Java Programming</LI><br><LI>Project</LI><br></UL><br></BODY><br></HTML><br>``` | **OUTPUT:-**<br><br>**D.I.T.-V**<br><br>&bull; WP with VB .Net<br>&bull; Wireless Communication<br>&bull; Dynamic Web Development<br>&bull; Java Programming<br>&bull; Project |

## <OL>:- (ORDERED LIST)

The OL tag creates an ordered list. The tag contains one or more LI tags that define the actual items of the list.

| Attribute | Value | Description |
|---|---|---|
| Type | 1,A,a,I,i | Specifies the style of the bullets of list items. |
| Start | Nnumber, Alphabet | Specifies the starting value of list. |

Value Attribute:-

1 – numbers (1,2,3...)
A – Uppercase letters (A,B,C....)
a – Lowercase letters (a,b,c...)
I – Uppercase Roman (I,II,III....)
i – Lowercase Roman (i,ii,iii...)

3.7.html (Ordered List)

| | |
|---|---|
| ```<br><HTML><br><BODY><br><H3>D.I.T.-V</H3><br><OL TYPE=1><br><LI>WP with VB .Net</LI><br><LI>Wireless Communication</LI><br><LI>Dynamic Web<br>Development</LI><br><LI>Java Programming</LI><br><LI>Project</LI><br></OL><BR><br><H3>D.I.T.-VI</H3><br><OL TYPE=1 START=6><br><LI>ASP .Net with VB .Net</LI><br><LI>Information Security</LI><br>``` | **OUTPUT:-** |

| <LI>Computer Logic Design</LI><br><LI>Programming in C#</LI><br><LI>Project</LI><br></OL><br></BODY><br></HTML> | **D.I.T.-V**<br><br>1. WP with VB .Net<br>2. Wireless Communication<br>3. Dynamic Web Development<br>4. Java Programming<br>5. Project<br><br>**D.I.T.-VI**<br><br>6. ASP .Net with VB .Net<br>7. Information Security<br>8. Computer Logic Design<br>9. Programming in C#<br>10. Project |
|---|---|

## 3.7 IMAGES:-

Sometimes we have some information in the form of images. in HTML, it is possible to design a web page containing some images as contents. Images are used to represent the information as well as to make the page attractive.

### <IMG>:- (IMAGE)

This tag is used to insert an image in web page. set its **src** attribute equal to the URL(file path) of the image. The URL can be either absolute URL or relative URL as discussed in chapter 1. The browser displays the image where the <IMG> tag occurs in the document. If you put an image tag between two paragraphs, the browser shows the first paragraph, then the image, and then the second paragraph.

| Attribute | Value | Description |
|---|---|---|
| Src | URL | Specifies the URL of an image. |
| Alt | Text | Specifies an alternate text for an image. |
| Align | Top,Bottom,Middle, Left,Right | Specifies the alignment of an image according to surrounding elements. |
| Border | Number(Pixels) | Specifies the width of the border around an image. |
| Width | Number(Pixels) | Specifies the width of an image. |
| Height | Number(Pixels) | Specifies the height of an image. |

3.8.html (Images)

| <HTML><BODY><br><H3>INDIA</H3><br><IMG SRC=INDIA.GIF WIDTH=200<br>HEIGHT=100 BORDER=3><BR><br></BODY></HTML> | INDIA<br><br>OUTPUT:- |
|---|---|

## 3.8 TABLES:-

A table places information inside the cells formed by dividing a rectangle into rows and columns. HTML supports the representation of information in tabular form using following tags.

### <TABLE>:-

This tag defines an HTML table. A simple HTML table consists of the TABLE tag and one or more TR, TH, and TD tags. The TR tag defines a row, the TH tag defines a header, and the TD tag defines a cell.

| Attribute | Value | Description |
|---|---|---|
| Align | Left, Center, Right | Specifies the alignment of a table according to surrounding text. |
| Bgcolor | Color | Specifies the background color of the table. |
| Background | Url | Specifies the background image of the table. |
| Border | Number(Pixels) | Specifies the width of the border of the table. |
| BorderColor | Color | Specifies the color of the border. |
| Cellpadding | Number(Pixels) | Specifies the spaces between the cell border and the cell content. |
| Cellspacing | Number(Pixels) | Specifies the spaces between cells. |

### <TR>:-

The <TR> tag defines a row in an HTML table. A TR tag contains one or more TH or TD tags.

| Attribute | Value | Description |
|---|---|---|
| Align | Left, Center, Right | Specifies the alignment of the content in a row. |
| Bgcolor | Color | Specifies the background color of the row. |
| Background | Url | Specifies the background image of the row. |

### <TD>:-

The <TD> tag defines a standard cell in an HTML table.
An HTML table has two kinds of cells:
Header cells -        contains header information (created with the TH tag)
Standard cells -    contains data (created with the TD tag)
The text in a TH tag is bold and centered.
The text in a TD tag is regular and left-aligned.

| Attribute | Value | Description |
|---|---|---|
| Align | Left, Center, Right | Specifies the alignment of the content in a cell. |
| Valign | Top,middle,bottom baseline | Specifies the vertical alignment of the content in a cell. |

| Bgcolor | Color | Specifies the background color of a cell. |
|---------|-------|-------------------------------------------|
| colspan | Number | Specifies the number of columns to merge. |
| Rowspan | Number | Specifies the number of rows to merge. |

3.9.html (Tables)

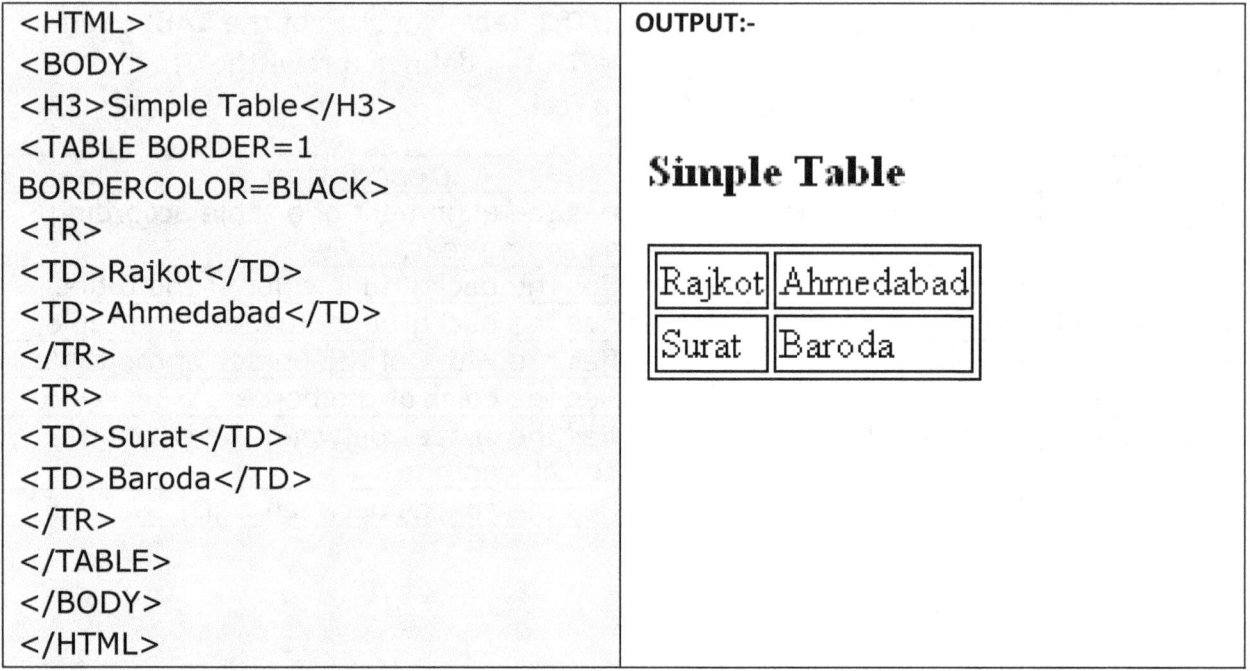

| <code>&lt;HTML&gt;<br>&lt;BODY&gt;<br>&lt;H3&gt;Simple Table&lt;/H3&gt;<br>&lt;TABLE BORDER=1<br>BORDERCOLOR=BLACK&gt;<br>&lt;TR&gt;<br>&lt;TD&gt;Rajkot&lt;/TD&gt;<br>&lt;TD&gt;Ahmedabad&lt;/TD&gt;<br>&lt;/TR&gt;<br>&lt;TR&gt;<br>&lt;TD&gt;Surat&lt;/TD&gt;<br>&lt;TD&gt;Baroda&lt;/TD&gt;<br>&lt;/TR&gt;<br>&lt;/TABLE&gt;<br>&lt;/BODY&gt;<br>&lt;/HTML&gt;</code> | OUTPUT:- |

**Description:-**

- In above example, we want to create a table with 2 columns and 2 rows. With border set to size 1 and its background color to black.
- First of all, we need to use <TABLE> tag with border and bordercolor attribute.
- Now, we want to add 2 rows and so we have to insert <TR> two times.
- In each row, we want to add 2 cells and so inside each <TR>, we have to insert two <TD>.
- As we discussed that a <TD> represents the exact content of a cell. So we have to write the content inside appropriate <TD> and its corresponding </TD>.
- It is always necessary to insert <TR> for new row and <TD> for new cell at proper place.
- If we start a new row by using <TR> without completing the previous row with </TR>, then also the table will look the same.

### 3.10.html (Tables)

```
<HTML>
<BODY>
<H3>D.I.T.-V</H3>
<TABLE BORDER=1 BORDERCOLOR=BLACK>
<TR>
<TH>CODE</TH><TH>SUBJECT NAME</TH><TH>CREDITS</TH>
</TR>
<TR>
<TD>350703</TD>
<TD>JAVA PROGRAMMING</TD>
<TD>7</TD>
</TR>
<TR>
<TD>351601</TD>
<TD>WINDOWS PROGRAMMING WITH VB.NET</TD>
<TD>8</TD>
</TR>
<TR>
<TD>351602</TD>
<TD>WIRELESS COMMUNICATION</TD>
<TD>5</TD>
</TR>
<TD>351603</TD>
<TD>DYNAMIC WEB DEVELOPMENT</TD>
<TD>7</TD>
</TR>
<TD>351604</TD>
<TD>PROJECT</TD>
<TD>6</TD>
</TR>
</TABLE></BODY></HTML>
```

**D.I.T.-V**

| CODE | SUBJECT NAME | CREDITS |
|---|---|---|
| 350703 | JAVA PROGRAMMING | 7 |
| 351601 | WINDOWS PROGRAMMING WITH VB.NET | 8 |
| 351602 | WIRELESS COMMUNICATION | 5 |
| 351603 | DYNAMIC WEB DEVELOPMENT | 7 |
| 351604 | PROJECT | 6 |

3.11.html (Tables)

| <HTML><br><BODY><br><TABLE BORDER=1<br>BORDERCOLOR=BLACK><br><TR><br><TD COLSPAN=2>DWD</TD><br><TD>JAVA</TD><br></TR><br><TR><br><TD>HTML</TD><br><TD>ASP</TD><br><TD>CORE JAVA</TD><br></TR><br></TABLE></BODY></HTML> | OUTPUT:-<br>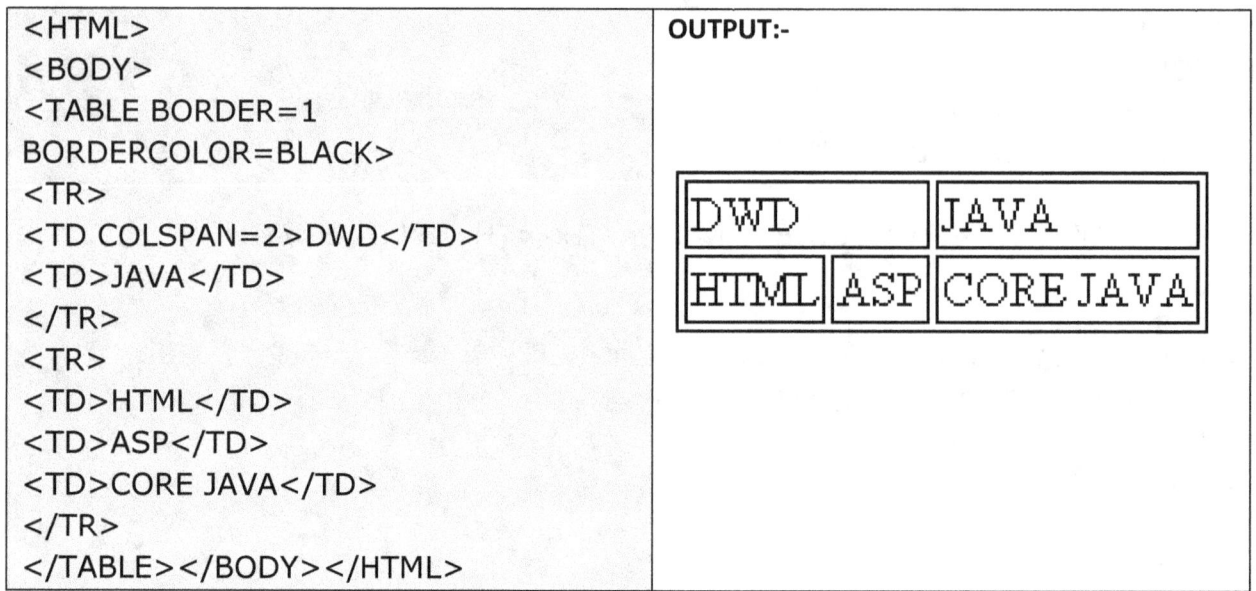 |
| --- | --- |

Description:-

Row-1:-

- In 1st row, we need only two cells and in 2nd row, we need three cells. So it is required to merge any two cells in the 1st row.
- The selection of the cells to merge depends on how we want to display the information in next row. In our example, we want to display "HTML" and "ASP" below the cell "DWD".

Row-2:-

- "HTML" is the 1st cell in 2nd row. And "ASP" is the 2nd cell in 2nd row. So we need to merge 1st cell and 2nd cell of 1st row to represent "DWD".
- As we are merging two cells by columns, we have to use COLSPAN=total number of columns, here it is 2, in the <TD> from where we want to start merging cells.
- Because we have only two cells in 1st row and three cells in 2nd row, we should include other <TR> and <TD> tags accordingly.

3.12.html (Tables)

| <HTML><br><BODY><br><TABLE BORDER=1<br>BORDERCOLOR=BLACK><br><TR><TD ROWSPAN=3>DWD</TD><br><TD>HTML</TD><br></TR><br><TR><br><TD>VB Script</TD> | OUTPUT:- |
| --- | --- |

```
</TR>
<TR>
<TD>ASP</TD>
</TR>
<TR>
<TD>JAVA</TD>
<TD>CORE JAVA</TD>
</TR>
</TABLE>
</BODY>
</HTML>
```

| DWD | HTML |
| | VB Script |
| | ASP |
| JAVA | CORE JAVA |

Description:-

Row 1:-

We need total 4 rows and 2 columns. To represent "DWD", we need to merge 3 cells (1st cell of three rows). So we have to use ROWSPAN=3 with the <TD> where we need to display "DWD"(1st cell of 1st row).

Now we can display "HTML" in 2nd cell of 1st row with simple <TD>. ]

Row 2:-

As we have already merged the 1st cell of 2nd row with the 1st cell of 1st row, we have to put a <TD> for the 2nd cell of 2nd row. In our example it is for "VB Script".

Row 3:-

As we have already merged the 1st cell of 3rd row with the 1st cell of 1st row, we have to put a <TD> for the 2nd cell of 3rd row. In our example it is for "ASP".

Row 4:-

Now we need two simple cells in 4th row. So we need to insert two <TD> tags. One is for "JAVA" and one is for "CORE JAVA".

## 3.9 HYPER LINKS:-

When we design a web site, we need to show lots of information on a single web site. Consider the web site of GTU. It contains information about syllabus, circulars, results, institutes etc.

43

If we keep all the information in a single web page, then it is very difficult to design as well as difficult for a user to find specific piece of information. So we distribute the information across multiple web pages. (Circulars, some pages for results etc.)

In a web page, we provide a set of links through which we can access other web pages or other documents. A link is a unidirectional pointer from one document that contains the link to some another document.

For linking purposes, we use the <a> anchor tag with its important attribute: href. The href attribute is set to the URL of the target document, which is the address of the document.

The <a> tag defines an anchor. An anchor can be used in two ways:

-To create a link to another document, by using the href attribute.

-To create a bookmark inside a document, by using the name attribute. Bookmarks are useful while creating a link to some another portion of the same document.

We need to provide some content (text/image) for the link. On the click event of that content, browser will open the corresponding document. This content is placed between <A...> and </A>

By default, links will appear as follows in all browsers:

An unvisited link is underlined and blue

A visited link is underlined and purple

An active link is underlined and red

| Attribute | Value | Description |
|---|---|---|
| Href | URL | Specifies the address of the document to link. |
| Name | Text | Specifies the name of an anchor. Works as a bookmark. |
| Target | _blank<br>_parent<br>_self<br>_top<br>framename | Specifies where to open the linked document. |

**Types of linking:-**

1- Linking to another document.
2- Linking inside the same document.
3- Linking to a specific portion of another document.
4- Email link.

## (1) Linking to another document.

3.13.html (Linking to another document)

| Code | Output |
|---|---|
| `<HTML>`<br>`<BODY>`<br>`<A HREF=3.10.html>DIT-V SYLLABUS</A>`<br>`</BODY>`<br>`</HTML>` | OUTPUT:-<br><br>DIT-V SYLLABUS |

We discussed 3.10.html file. In above program, when user will click on

"DIT-V SYLLABUS" text. Browser will open the 3.10.html file. This is the linking of 3.10.html from 3.13.html.

3.14.html (Linking to another web site)

| Code | Output |
|---|---|
| `<HTML>`<br>`<BODY>`<br>`<A HREF="http://www.yahoo.com">Yahoo!</A><BR>`<br>`<A HREF="http://www.google.co.in">Google</A><BR>`<br>`<A HREF="http://www.hotmail.com">HotMail</A><BR>`<br>`</BODY>`<br>`</HTML>` | OUTPUT:-<br><br>Yahoo!<br>Google<br>HotMail |

3.15.html (Linking using an image)

| Code | Output |
|---|---|
| `<HTML>`<br>`<BODY>`<br>`<A HREF="http://www.google.com">`<br>`<IMG SRC="google.png">`<br>`</A>`<br>`</BODY>`<br>`</HTML>` | OUTPUT:-<br><br>Google |

It is also possible to use an image instead of simple text. When user will click on the image of nirav, browser will open the http://www.niravprakasan.com.

## (2) Linking inside the same document.

45

Sometimes we want to put lots of information in a single web page. At this time, if we don't provide linking facility inside the same web page, then it will be difficult for the users to find the information easily. Users will need to use scroll bars to move from one portion to another portion of content inside the same web page.

Consider an example of the syllabus. We want to display the entire syllabus of Diploma I.T. engineering. One way is to use a page from where we can open the syllabus of individual semesters as explained in Example 3.13

One another possibility is to put the syllabuses of all the semesters in a single page as explained in Example 3.7 Provide various links at the top of the page for all the semesters; we can do the linking in such a way that, when user will click on a specific link of a specific semester, browser will display the content of that semester.

This type of linking is the linking inside the same document because we don't have separate pages. Now to identify the beginning of the syllabus of various semesters, we need to mark the starting points of the syllabus of various semesters. For this purpose we use the <A> tag with its name attribute.

<A NAME="sem1"></A> Wherever we will write this tag, browser will identify that location as "sem1" so later on we can reach there by specifying the URL of the current page as well as the location which is "sem1" in <A>.

3.16.html (Linking in same document)

| <HTML> <BODY> <A NAME="top"> <A HREF="#bottom">BOTTOM OF THE PAGE</A> <BR><BR><BR><BR><BR><BR> <BR><BR><BR><BR><BR><BR><BR> BR><BR><BR><BR><BR><BR><BR> <A NAME="bottom"> <A HREF="#top">TOP OF THE PAGE</A> </BODY> </HTML> | OUTPUT:- <br><br>BOTTOM OF THE PAGE <br><br><br><br>TOP OF THE PAGE |
|---|---|

In above example, two bookmarks are placed at two different positions(one is at the top of the page "top" and one is at the bottom of the page "bottom"). We can give the address of a bookmark using #bookmarkname.

When user will click on "BOTTOM OF THE PAGE", browser will open the portion of the same document which is marked as "bottom" using <A NAME="bottom">.

### (3) Linking to a specific portion of another document.

Suppose we have two files 3.16.html (which contains two locations marked as "top" and "bottom") and 3.17.html. we want to insert the same hyperlinks similar to the example 3.16 but from different file.

<A HREF=filename#bookmarkname>TEXT</A>. This is the tag which refers to a link to a specific portion of another document.

3.17.html (Linking to a specific portion of another document)

| | OUTPUT:- |
|---|---|
| ```<HTML><BODY><A NAME="top"><A HREF="3.16.html#bottom">BOTTOM OF THE 3.16.HTML</A><BR><A NAME="bottom"><A HREF="3.16.html#top">TOP OF THE 3.16.HTML</A></BODY></HTML>``` | BOTTOM OF THE 3.16.HTML<br>TOP OF THE 3.16.HTML |

In above example, when user will click on "BOTTOM OF THE 3.16.HTML", browser will open the 3.16.html and because the complete URL is not only 3.16.html but 3.16.html#bottom, so it will directly show the portion of the 3.16.html which is marked as "bottom".

### (4) Email link.

If we want users to send us ab email from the Web page, we can use the "mailto" option in HREF attribute of <A>. When a user will click on the link, the browser will open the default mail client of user's computer to allow him to send us an email.

47

3.18.html (Email Link)

| | OUTPUT:- |
|---|---|
| `<HTML>`<br>`<BODY>`<br>`<A HREF="mailto:hardik.molia@gmail.com">Mail Me</A>`<br>`</BODY>`<br>`</HTML>` | Mail Me |

## 3.10 FRAMES:-

A framed page divides a browser window into multiple paneLGs, or smaller window frames. Each frame can contain a different page. The benefit of frames is that users can view information in one frame while keeping another frame open. The contents of one frame can be manipulated, or linked, to the contents of another. For example, one frame can contain links that displays a specific page in another frame.

### <FRAMESET>:-

This tag is used to create a framed page. This tag is the replacement of the <BODY> tag in a page containing frames.

A frameset hold one or more frames. And each frame holds one page. It is possible to have more than framesets in a single framed page.

The frameset element specifies total number of columns or rows there will be in the frameset, and the amount of space(percentage/pixels) will occupy by each of the frames.

| Attribute | Value | Description |
|---|---|---|
| cols | Pixels<br>%<br>* | Specifies the number and size of columns in a frameset. |
| rows | Pixels<br>%<br>* | Specifies the number and size of rows in a frameset. |

Value:-    There are three ways to specify the size of columns/rows of frames inside a frameset.

**(1)    pixels:-**
cols = "300,200" means that we want two frames in two columns. 1st(left frame) is of 300 pixel and 2nd(right frame) is of 200 pixel. This method requires the knowledge about the resolution of the client system.

**(2)    %:-**
Instead of giving the values in pixels absolutely, we can specify the size in percentage with respect to the screen resolution. If we want to partition the page with exactly two same columns, We can write cols="50%, 50%"

**(3)    *:-**

It is also possible to set one frame with a fixed size without taking care about the screen resolution. Suppose we want to partition the page with two columns with left most column size is 300. our requirement is that the right most column size must be set to the width of the page – 300 automatically. We can use * with the cols attribute.

Cols = "300,*" where * indicates remaining space of the screen.

## <FRAME>:-

This tag is used to define one particular frame within a frameset.

| Attribute | Value | Description |
|---|---|---|
| Src | URL | Specifies the URL of the document to show in a frame. |
| Name | Name | Specifies the name of a frame. |
| Noresize | Noresize | Specifies that a frame cannot be resized. |
| Scrolling | Yes,No,Auto | Specifies whether or not to display scrollbars in a frame. |

3.19.html (Frames – two columns)

| <HTML> | OUTPUT:- |
|---|---|
| <HTML><br><FRAMESET COLS="100,*"><br><FRAME SRC=3.14.html BORDER=1><br><FRAME SRC=3.13.html BORDER=1><br></FRAMESET><br></HTML> | Yahoo!<br>Google<br>HotMail        DIT-V SYLLABUS |

## Description:-

We have already discussed 3.14.html and 3.13.html in examples 3.14 and 3.13 respectively. Here we want to display both the files in a single web page 3.19.html. So we need to create a <FRAMESET> with 100 pixels for the left most column and * to indicate the rest of the pixels for the right most column.

In between of <FRAMESET> and </FRAMESET>, we have to specify which web pages, we want to show inside the frames using <FRAME> tag.

3.20.html (Frames – two columns)

| | |
|---|---|
| `<HTML>`<br>`<FRAMESET COLS="20%,80%">`<br>`<FRAME SRC=3.14.html BORDER=1>`<br>`<FRAME SRC=3.13.html BORDER=1>`<br>`</FRAMESET>`<br>`</HTML>` | OUTPUT:-<br><br>Yahoo!　｜　DIT-V SYLLABUS<br>Google<br>HotMail |

It is also possible to give the amount of spaces for the frames in percentage instead of absolute pixels as shown in above example.

`<FRAMESET COLS="20%,*">` is a valid syantax

`<FRAMESET COLS="20%,50%">` is a valid syntax. Because 20% + 50% = 70%. The remaining 30% portion of the web page will remain empty.

3.21.html (Frames – two rows)

| | |
|---|---|
| `<HTML>`<br>`<FRAMESET ROWS="20%,80%">`<br>`<FRAME SRC=3.14.html BORDER=1>`<br>`<FRAME SRC=3.13.html BORDER=1>`<br>`</FRAMESET>`<br>`</HTML>` | OUTPUT:-<br><br>Yahoo!<br>Google<br>HotMail<br><br><br>DIT-V SYLLABUS |

In above program, we have put two frames in two rows using the ROWS attribute. All other rules of the specifying the sizes are same.

3.22.html (Frames – three rows)

| | |
|---|---|
| `<HTML>`<br>`<FRAMESET ROWS="20%,20%,*">`<br>`<FRAME SRC=3.14.html BORDER=1>`<br>`<FRAME SRC=3.13.html BORDER=1>`<br>`<FRAME SRC=3.18.html BORDER=1>`<br>`</FRAMESET>`<br>`</HTML>` | OUTPUT:-<br><br>Yahoo!<br>Google<br>HotMail<br><br>DIT-V SYLLABUS<br><br>Mail Me |

3.23.html (Frames – Nesting)

| | OUTPUT:- |
|---|---|
| `<HTML>`<br>`<FRAMESET ROWS="80%,*">`<br>`<FRAMESET COLS="20%,80%">`<br>`<FRAME SRC=3.14.html BORDER=1>`<br>`<FRAME SRC=3.13.html BORDER=1>`<br>`</FRAMESET>`<br>`<FRAME SRC=3.18.html BORDER=1>`<br>`</FRAMESET>`<br>`</HTML>` | Yahoo!<br>Google<br>HotMail  DIT-V SYLLABUS<br><br>Mail Me |

A single <FRAMESET> can divide the page either row wise or column wise. So if we want to have a combination of columns and rows as shown in above example, we need to use more than one <FRAMESET> tags.

In above example, we want to display three files. First of all we need to divide the page into two rows. 1st row will be used to display 3.14.html and 3.13.html. and 2nd row will be used to display 3.18.html

<FRAMESET ROWS="80%,*"> will divide the page into two rows. Now according to our requirement we need to divide the 1st row into two columns. So <FRAMESET COLS="20%,80%"> will divide the 80% of 1st row into two columns. So now we have three place holders in which we can assign different frames.

As we are inside the area of <FRAMESET COLS="20%,80%">, we should specify the web pages corresponding to 1st row(left column of 20% and right column of 80%) as shown below.

<FRAME SRC=3.14.html BORDER=1>

<FRAME SRC=3.13.html BORDER=1>

Now this is the end of the 1st row. But actually 1st row is a set of 2 columns created using a <FRAMESET>. So before moving to the 2nd row, we need to specify the end of the <FRAMESET>, which is representing two columns inside the 1st row. So we should use </.FRAMESET> to indicate the ending of the divisions inside the 1st row. Now we can specify the web page for the 2nd row using <FRAME SRC=3.18.html BORDER=1>

At last we use </.FRAMESET> to indicate the ending of the frameset containing both the rows.

3.24.html (Frames – Target to a frame)

| | |
|---|---|
| ```html<br><HTML><br><FRAMESET ROWS="10%,*"><br><FRAME SRC=3.24.1.html<br>BORDER=1><br><FRAME NAME=f1 BORDER=1><br></FRAMESET><br></HTML><br>``` | OUTPUT:- 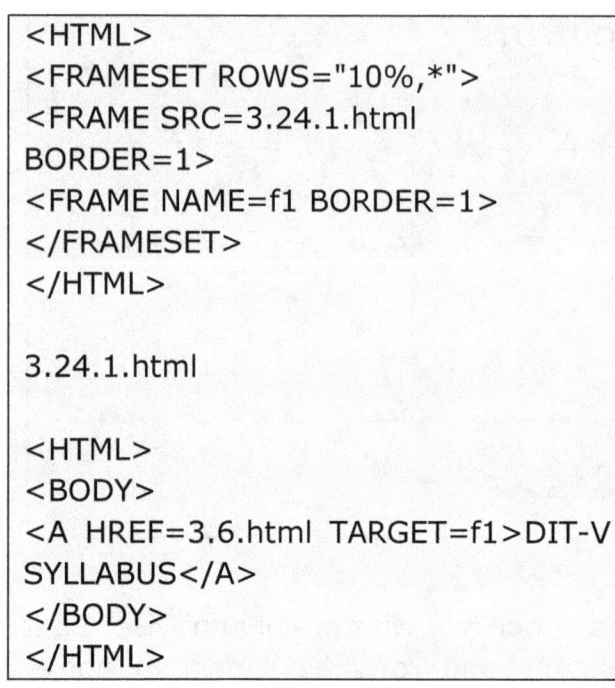 |

3.24.1.html

```html
<HTML>
<BODY>
<A HREF=3.6.html TARGET=f1>DIT-V
SYLLABUS</A>
</BODY>
</HTML>
```

Suppose we want to design a page as shown above. When user will click on "DIT-V SYLLABUS" from one frame, browser should open the page containing the syllabus in another frame.

For this purpose we need to use linking as shown in 3.24.1.html. because we need to display 3.6.html fileinside a frame, we have to set <A>'s target attribute to the frame name. here it is "f1"

Now we also need to name the frame inside the <FRAME> tag. So now 3.24.html contains two frames. 1st (top most) frame displays 3.24.1.html containing the link to 3.6.html targeting to the frame "f1". Browser will display the content of 3.6.html in the frame which is named as "f1" when user will click on the "DIT-V SYLLABUS" because of the TARGET=f1.

3.25.html (Frames – Inline frame)

| | |
|---|---|
| ```html<br><HTML><br><BODY><br><IFRAME SRC=3.6.html BORDER=1<br>WIDTH=200 HEIGHT=100><br></HTML><br>``` | 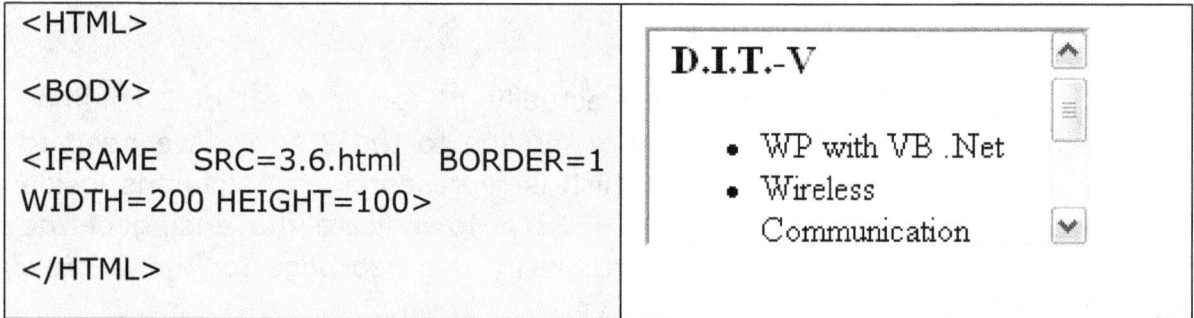 |

It is also possible to insert a web page in another web page just like an image or a table. This type of insertion is possible using <IFRAME> inline frame. We can set the width and height properties and accordingly scrollbars will be available to navigate a page inside the inline frame.

## 3.11 FORMS:-

While links provide basic ability for users to make choices for navigating a web site, fill-out forms enable users to submit information that can be used to create an interactive environment which is the basic need of dynamically created Web site. A form is a collection of controls like textbox, radio button, check box, command button, list box etc. in this section we will study how to create a form. We will study more about the forms in chapter 6

### <FORM>:-

This tag is used to create form. It also specifies how and where client browser will send the information filled in the various fields of the form.

| Attribute | Value | Description |
|-----------|-------|-------------|
| Name | Name | Specifies the name for a form. |
| Method | Get,Post | Specifies how to send form-data to the server. |
| Action | URL | Specifies where to send the form-data when a form is submitted. |

### <INPUT>:-

This tag is used to create various controls to get information.

| Attribute | Value | Description |
|-----------|-------|-------------|
| Name | Name | Specifies a name for an input element. |
| Type | Button,Checkbox File,Hidden,Image Password,Radio Reset,Submit,Text | Specifies the type of an input element. |
| Maxlength | Nnumber | Specifies the maximum length (in characters) of an input field (for type="text" or type="password") |
| Readonly | Readonly | Specifies that an input field should be read-only (for type="text" or type="password") |
| Checked | Checked | Specifies that an input field should be already selected when the page loads (for type="checkbox" or type="radio") |

## <TEXTAREA>:-

The <TEXTAREA> tag defines a multi-line text input control. The size of a text area can be specified by the cols and rows attributes.

| Attribute | Value | Description |
|-----------|-------|-------------|
| Name | Name | Specifies a name for a textarea. |
| Cols | Number | Specifies the visible width of a textarea. |
| Rows | Number | Specifies the visible height of a textarea. |

It is also possible to set the value of a text area at the design time by using following syntax. <TEXTAREA> value </TEXTAREA>.

## <SELECT>:-

The <SELECT> tag is used to create a select list (drop-down list and list box).The <OPTION> tags inside the select element define the available options in the list.

| Attribute | Value | Description |
|-----------|-------|-------------|
| Name | Name | Specifies a name for a select list. |
| multiple | Multiple | Specifies that multiple options can be selected. |
| Size | Number | Specifies the number of visible options in a list. If this attribute is set then a list box is created otherwise a drop down list is created. |

3.26.html (Form)

```
<HTML>
<BODY>
<FORM>

NAME:- <INPUT TYPE="TEXT" NAME="T1"></INPUT><BR><BR>

PASSWORD:- <INPUT TYPE="PASSWORD" NAME="P1"></INPUT><BR><BR>

DIPLOMA:- <INPUT TYPE="CHECKBOX" NAME="C1"></INPUT><BR><BR>

DEGREE:- <INPUT TYPE="CHECKBOX" NAME="C2"></INPUT><BR><BR>

MALE <INPUT TYPE="RADIO" NAME="R1"></INPUT>
FEMALE<INPUT TYPE="RADIO" NAME="R1"></INPUT><BR><BR>

ADDRESS:- <TEXTAREA NAME="T2" ROWS=5
COLS=25></TEXTAREA><BR><BR>
```

```
BRANCH:- <SELECT NAME="B1">
<OPTION>CIVIL
<OPTION>MECH.
<OPTION>I.T.
<OPTION>E.C.</SELECT>
<BR><BR>

CELL CONNECTION:-
<SELECT NAME="CN1" SIZE=4 MULTIPLE>
<OPTION>!DEA
<OPTION>VODAPHONE
<OPTION>AIRTEL
<OPTION>BSNL</SELECT><BR><BR>

<INPUT TYPE = "SUBMIT">
<INPUT TYPE = "RESET">
<INPUT TYPE = "BUTTON" VALUE="Exit">

</FORM>
</BODY>
</HTML>
```

| OUTPUT:- | |
|---|---|
| | We have two check boxes C1 (diploma) and C2 (degree). Logically it is correct to select either of the both or both of them or none of them at a time and so we gave two different names to both of them. |
| | We have two radio buttons one is for the male and another is for the female. Now it is not required to select both of them at a time. And so we gave both the radio buttons a same name R1. so user will able to select only one of them. |
| | B1 is a drop down list for the branch selection. CN1 is a selection list for the cell phone |

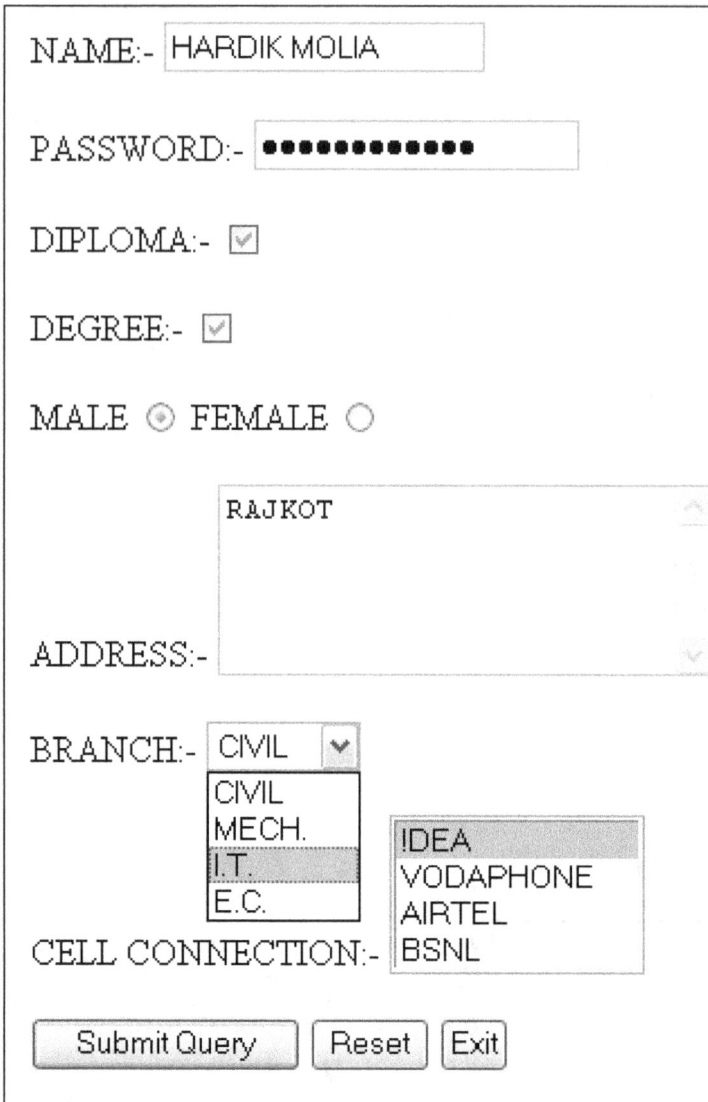

connection. We have set size attribute of CN1 and so browser will create a list box instead of a default drop down list.

There are three buttons. The submit button sends the form data to the page using the method specified by the action and method attributes of the <FORM>.

The reset button clears all the information entered into form controls.

The Exit button is the user defined button. User can program its click event.

| Field Name | Type | Description |
|---|---|---|
| T1 | TEXT | A simple single lined textbox. |
| P1 | PASSWORD | A password field. |
| C1 | CHECKBOX | A checkbox. |
| C2 | CHECKBOX | A checkbox. |
| R1 | RADIO | A radio button for male. |
| R1 | RADIO | A Radio button for female. |
| T2 | TEXTAREA | A multi line textbox. |
| B1 | SELECT | A drip down list. |
| CN1 | SELECT | A list box. |

## 3.12 MISCELLANEOUS TAGS:-

### <BODY>:-

This tag is used to define the page's properties like background image, background color, default text color etc. this tag contains all the content of a page such as text, links, images, lists, tables, forms etc.

| Attribute | Value | Description |
|---|---|---|
| Background | URL | Specifies a background image. |
| Bgcolor | Color | Specifies a background color. |
| Text | Color | Specifies default text color. |

<BODY BGCOLOR=BLACK>

This will set the background color of the page to black.

### <HR>:-

This tag is used to create a horizontal line in a page. This tag is used to separate the contents.

| Attribute | Value | Description |
|---|---|---|
| Align | Left,Center,Right | Specifies the alignment of the line. |
| Size | Pixels | Specifies the height of the line. |
| Width | Pixels | Specifies the width of the line. |
| Color | Color | Specifies the color of the line. |

<HR COLOR=RED>

This will create a horizontal red line at the place where we write this tag in page.

### <EMBED>:-

This tag is used to insert a browser plug-in in the page. A plug-in is a special program located on the client computer that handles its own special types of files. The most common plug-ins are for movies and sounds.

| Attribute | Value | Description |
|---|---|---|
| Src | URL | Specifies the location of the file. Movie file, flash .swf file etc. |
| Autostart | True,false | Specifies whether the movie should auto start with the page load or not. |
| Width | Pixels | Specifies the width of the plug-in in page. Width |

| | | |
|---|---|---|
| | | of the player for a movie. |
| Height | Pixels | Specifies the height of the plug-in in page. Height of the player for a movie. |
| Loop | Number | Specifies the number of times the movie should repeat itself. |

## <MARQUEE>:-

This tag is used to create a scrolling marquee (i.e. scrolling text or scrolling images) by using the <MARQUEE> tag. You can make the text/images scroll from right to left, left to right, top to bottom, or bottom to top. The content which we want to scroll is placed between <MARQUEE> and </MARQUEE>.

| Attribute | Value | Description |
|---|---|---|
| Behavior | Scroll (Default) , Slide, Alternate | Specifies the behavior of the marquee. How text scrolls. |
| Direction | Left(Default), Right, Up, Down | Specifies the direction of the marquee. Default is from right to left. |
| ScrollAmount | Number | Specifies the amount in pixels the marquee should scroll each time it's redrawn. The default value is 6. |
| ScrollDelay | Number | Specifies the amount in milliseconds between times the marquee is redrawn. The default value is 85. |
| Loop | Number | Specifies the number of times the marquee should repeat itself when it reaches the end of the message. The default is "infinite". |

<MARQUEE BEHAVIOR=ALTERNATE>Asp Programming</MARQUEE>

This will create a scrolling marquee with alternate behavior for the text "Asp Programming"

<MARQUEE BEHAVIOR=SLIDE><IMG SCR="1.jpg"></MARQUEE>

This will create a scrolling marquee with slide behavior for the image 1.jpg.

## <APPLET>:-

This tag tells the browser to load the applet whose Applet subclass is specified in code attribute and display it I an area of specified width and height.

| Attribute | Value | Description |
|-----------|-------|-------------|
| Code | URL | Specifies the location of the .class file of java applet. |
| Width | Pixels | Specifies the width of the applet in page. |
| Height | Pixels | Specifies the height of the applet in page. |

We can pass parameters to the Applet using <PARAM> tag with its name and value attributes.

<APPLET CODE=myapplet.class WIDTH=200 HEIGHT=200><PARAM NAME=city VALUE=Rajkot></APPLET>

## <STYLE>:-

This tag is used to define style information for an HTML document. Inside the style tag we define how elements should display by the browser. This tag must be placed inside the <HEAD>.we will study this tag in detail in next topic.

## <SCRIPT>:-

This tag is used to insert client side script (java script) which browser executes on client machine. This tag must be placed inside the <HEAD>.we will study this tag in detail in chapter 4

## 3.13 CASCADING STYLE SHEETS:-

Cascading Style Sheets (CSS) is a language used to embed style sheets inside a web page. It is used to describe the formatting information of a tab.

With HTML only, we write the actual information and information about its formatting together inside the body element.

Style sheet is a set of statements (separated from the actual content) which is used to specify how various elements of a page will be displayed by the browser.

Imagine that you want to display each table cell's content in blue. So instead of modifying every <TD> of a table, using CSS we can set the style of <TD> only once in a style sheet and that will apply by the browser to all <TD> tags.

Suppose you want to set the border of all the images of a page to pixel 2. Instead of write border=2 in each <IMG>, using CSS we can set the style only once and that will apply by the browser to all <IMG> tags.

### 3.13.1    Advantages:-

### (1) Various formatting options.

CSS supports various easy to use formatting options. Some options are very difficult to implement in HTML but very easy to use using CSS.

### (2) Less code to type.

In CSS, we don't need to write the formatting with every use of a tag. We write the formatting only once, and it is applicable through out the page(internal style sheets) and web site(external style sheet).

In HTML, we need to use specific formatting tags wherever we want a specific effect. Suppose if we want to set text color of a paragraph text in alternate fashion.(one paragraph in red, next is in blue, next is in red....) then at the starting of each paragraph, we need to set the color to either red or blue. In CSS, we can do this type of formatting very easily using classes.

### (3) Maintenance and error handling is also easy.

In HTML we write the code for the formatting wherever we want. So we need to write the same code more than once. So if in future, we want to modify the formatting, we will need to change all the occurrences of that formatting code which is time consuming task.

In CSS, we write the formatting code only once, and browser applies it to all the elements of the browser. So it is very easy to modify the formatting by changing the format in CSS.

### (4) Apply changes globally.

We often need to maintain the same format at page level as well as at web site level. CSS supports the concept of external style sheets. External style sheets are special .css files, which store style sheets. We import or link these files to our web pages and formatting code is ready to use.

Suppose we want to design a website of 10 web pages. We want to have a common paragraph format in all the pages. Suppose the format is left alignment, verdana font and blue color.

We can specify the format with each paragraph in each of the 10 pages using HTML tags. This is quite inefficient way.

Another way is to specify the formatting code in every page using an internal style sheet. But in this option, we need to write the same code 10 times.

The best way is to store the formatting code as an external style sheet and link/import it in all the 10 pages. This is the best way to apply the formatting globally.

## (5) Change in basic formatting associated with tags.

When we use <P>........</P>, browser will display the paragraph with its default formatting. We can change the default formatting of any tag using style sheets.

### 3.13.2 Types of Style Sheets:-

There are four ways we can use the style sheets according to the place where we store the style information.

(1) Internal style sheets:-

(1.1) Embedding style sheet

We write the style definitions at the page level. We use <STYLE> tag inside the <HEAD> section. The style definitions are applicable in that page.

(1.2) Inline style sheet

In inline style sheets, we use the style attribute with the tag of which we want to change the format. The style definitions are applicable in that tag only.

(2) External style sheets:-

External style sheets are used to maintain the same style information across multiple pages. We store the style information in a separate file with .css extention.

(2.1) Linked style sheet

We can link a .css file using <LINK> tag.

(2.2) Imported style sheet

We can import a .css file using @import statement.

### 3.13.3 Syntax of Style Sheets:-

The syntax of writing a style sheet is different than of standard HTML tag. But it is easy to understand with reference with the syntax of a HTML tag.

The syntax of a HTML tag is:-
<TagName AttributeName1=value1 AttributeName2=value2>...</TagName>

The syntax of a style sheet selector is:-

SelectorName {property1:value1;property2:value2;property3:value3}
SelectorName is similar to the TagName.

PropertyName is similar to the AttributeName.

Value of a property is similar to the value of an attribute.

### Units:-

It is very necessary to give the measurement properly. A style sheet supports various units as given below.

### Absolute Length Units:-

%, Cm – centimeter, in – inch, mm – millimeters,
pc(picas) = 1/6 of an inch , pt(points) = 1/72 of an inch so  1 pc = 12 pt

### Relative Length Units:-

Px – A dot on a computer screen
Em – 1em is equal to the width of the 'M' of the currently used font. 2em is double.
Ex – 1ex is equal to the height of the 'x' of the currently used font.

### Colors:-

We can represent color information in various ways.

Color name like red,green,blue etc.

 rgb(redval, greenval, blueval) or rgb(%,%,%) using the rgb function.

#RRGGBB where RR,GG and BB are the red, green and blue components in Hexadecimal.

### URL:-

To specify a URL as a property value, we use URL('url') function.

## 3.13.4 Style Sheet Properties:-

**Text Properties:-**

| | |
|---|---|
| 1. Color | color |
| 2. Text-Direction | ltr or rtl (IE 6.0) |
| 3. Text-Align | left, right, center, justify |
| 4. Text-Decoration | underline, overline, line-through, blink |
| 5. Text-Transform | capitalize, uppercase, lowercase |
| 6. Letter-Spacing | number, %age |
| 7. Word-Spacing | number, %age |
| 8. Text-Indent | lenghth,%age |
| 9. Text-Shadow | value |

**Font Properties:-**

| | |
|---|---|
| 1. Font-Family | font name |
| 2. Font-Size | small, large, medium, size |
| 3. Font-Weight | bold, lighter |
| 4. Font-Style | italic, oblique |

**Background Properties:- (Can be used for other elements also)**

| | |
|---|---|
| 1. Background-Attachment | scroll, fixed |
| 2. Background-Color | color |
| 3. Background-Image | URL |
| 4. Background-Position | top left, top center, top right, center left, center center, center right, bottom left, bottom center, bottom right,   X% and Y%, Xpos and Ypos |
| 5. Background-Repeat | no-repeat, repeat-x, repeat-y |

**Margin Properties:- (Sets margin at page level)**

| | |
|---|---|
| 1. Margin-Left | Amount in pixels |
| 2. Margin-Right | Amount in pixels |
| 3. Margin-Top | Amount in pixels |

4. Margin-Bottom                Amount in pixels

## Padding Properties:- (Sets margin at element level)

1. Padding-Left           Amount in pixels

2. Padding-Right         Amount in pixels

3. Padding-Top            Amount in pixels

4. Padding-Bottom       Amount in pixels

## Border Properties (Can be used with almost all elements)

1. Border                  width style color

2. Border-Left            width style color

3. Border-Right          width style color

4. Border-Top            width style color

5. Border-Bottom        width style color

6. Border-Color          color

7. Border-Style           hidden, dotted, dashed, solid, double, ridge, inset, outset

8. Border-Width          thin medium thick length

9. Border-Left-Color     color

10. Border-Left-Style    hidden, dotted, dashed, solid, double, ridge, inset, outset

11. Border-Left-Width   thin medium thick length

## List Properties

1. List-Style             type position image

2. List-Style-Image     url(URL)

3. List-Style-Position  inside, outside (indentation difference)

4. List-Style-Type      disc, circle, square, decimal, lower-roman, upper-roman, lower-alpha, upper-alpha

5. Counter-Increment  +ve, 0 or -ve

## Positioning

| | |
|---|---|
| 1. Static | Used by HTML |
| 2. Absolute | Based on the browser window |
| 3. Relative | Relative positioning uses the same four positioning properties as absolute positioning. But instead of basing the position of the element upon the browser view port, it starts from where the element would be if it were still in the normal flow. |
| 4. Fixed | Similar to absolute but element doesn't move when page is scrolled. |

## 3.13.5 Style Sheet Examples:-

## (1) Internal Style Sheet:-

### 1.1 Embedding a style sheet in <HEAD> section using <STYLE> tag:-

3.27.html (TEXT and FONT PROPERTIES)

```
<HTML>
<HEAD>
<STYLE>

P{COLOR:red;
TEXT-DECORATION:UNDERLINE;
TEXT-TRANSFORM:CAPITALIZE;
LETTER-SPACING:4;}

H1{FONT-FACE:VERDANA;
FONT-SIZE:1CM;}

H2{TEXT-DECORATION:OVERLINE;FONT-
STYLE:ITALICS;COLOR:BLUE;WORD-SPACING:2CM;}

</STYLE>
</HEAD>
<BODY>
<P>Text Properties</p><BR>
<H1>H1 with Font Properties</H1><BR>
<H2>H2 with Text and Font Properties</H2>
</HEAD>
```

```
</HTML>
```

OUTPUT:-

Text Properties

# H1 with Font Properties

H2         with          Text          and
Font          Properties

3.28.html (BACKGROUND PROPERTIES)

```
<HTML>
<HEAD>
<STYLE>
BODY{BACKGROUND-IMAGE:URL("MAP.JPG");BACKGROUND-REPEAT:NO-
REPEAT}
P{COLOR:WHITE;TEXT-DECORATION:UNDERLINE;TEXT-
TRANSFORM:CAPITALIZE;
LETTER-SPACING:4;BACKGROUND-COLOR:GREEN;}
</STYLE>
</HEAD>
<BODY>
<P> This is text with underline with capitalize effect of text transformation.
</P><BR>
</BODY>
</HTML>
```

OUTPUT:-

This Is Text With Underline With
Capitalize Effect Of Text
Transformation.

## 3.29.html (BORDER PROPERTIES)

```
<HTML>
<HEAD>
<STYLE>
P{COLOR:RED;TEXT-DECORATION:UNDERLINE;TEXT-
TRANSFORM:CAPITALIZE; LETTER-SPACING:4;BORDER:3 DASHED;}
H5{BORDER-BOTTOM-STYLE:DOTTED;}
</STYLE>
</HEAD>
<BODY>
<P> This is text with dashed border</P><BR>
<H5>This is text with dotted border as an underline</H5><BR>
</HEAD>
</HTML>


OUTPUT:-



This Is Text With Dashed Border


This is text with dotted border as an underline
..............................................................
```

## 3.30.html (POSITION PROPERTIES)

```
<HTML>
<HEAD>
<STYLE>
P {POSITION:ABSOLUTE;LEFT:10;TOP:50;FONT-SIZE:1CM;}
</STYLE>
</HEAD>
<BODY>
<IMG SRC="MAP.JPG" WIDTH=200 HEIGHT=200>
<P> THIS IS TEXT ON AN IMAGE.</P>
</BODY>
</HTML>
```

**OUTPUT:-**

3.31.html (Z-INDEX)

```
<HTML>
<HEAD>
<STYLE>
H1{POSITION:ABSOLUTE;TOP:100;LEFT:100;WIDTH:300;HEIGHT:200;BAC
KGROUND-IMAGE:URL("MAP.JPG");Z-INDEX:1;}

H2{POSITION:ABSOLUTE;TOP:210;LEFT:50;WIDTH:300;HEIGHT:50;BACKG
ROUND-IMAGE:URL("WINTER.JPG");Z-INDEX:1;COLOR:WHITE;}
</STYLE>
</HEAD>
<BODY>
<H1> THIS IS LAYER 1 WITH ZINDEX 1 AND SUNSET BACKGROUND</H1>
<H2> THIS IS LAYER 2 WITH ZINDEX 2</H2>
</BODY></HTML>
OUTPUT:-
```

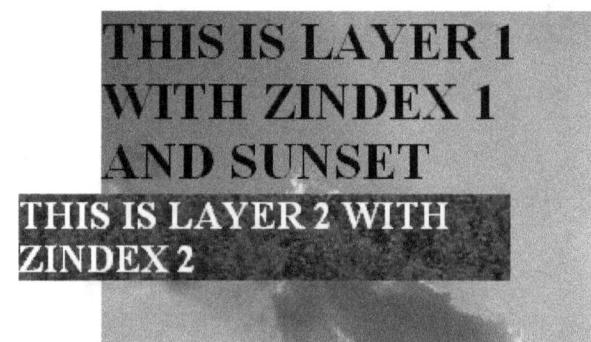

## Inline Style Sheet using style attribute:-

3.32.html (INLINE STYLE SHEET)

```
<HTML>
<BODY>
<P STYLE=
"TEXT-DECORATION:UNDERLINE;FONT-FACE:VERDANA;
BACKGROUND-COLOR:YELLOW;">
This is text with underline and other effects using stylesheet.</P>
<H1 STYLE="TEXT-DECORATION:OVERLINE;TEXT-
TRANSFORM:LOWERCASE;">
FONT EFFECT</H1>
</BODY>
</HTML>


OUTPUT:-


This is text with underline and other effects using stylesheet.

font effect
```

## (2) External Style Sheet:-

In external style sheet, we store the style definitions in a separate .css file. Here it is a.css

```
H1 {BACKGROUND-COLOR:BLUE;TEXT-DECORATION:OVERLINE}

H2{BACKGROUND-COLOR:BLACK;TEXT-TRANSFORM:UPPERCASE}
```

## 2.1 linking a style sheet using <LINK>:-

3.33.html (INLINE STYLE SHEET)

```
<HTML>
<HEAD>
<LINK REL="STYLESHEET" HREF="A.CSS">
</HEAD>
<BODY>
<H1>EFFECT WITH H1</H1><BR>
<H2>EFFECT WITH H2</H2>
```

```
</HEAD>
</HTML>
OUTPUT:-
```

**EFFECT WITH H1**

EFFECT WITH H2

## 2.2 Importing a style sheet using @import:-

3.34.html (INLINE STYLE SHEET)

```
<HTML>
<HEAD>
<HEAD>
<STYLE>
@import url("A.CS");
</STYLE>
</HEAD>
<BODY>
<H1>EFFECT WITH H1</H1><BR>
<H2>EFFECT WITH H2</H2>
</HEAD>
</HTML>


OUTPUT:-
```

**EFFECT WITH H1**

EFFECT WITH H2

It is always possible to override the style at next level. Inline style sheets override the style information of embedded style information.

For example, if we mention red color for <H1> inside the <STYLE> tag and again blue color for <H1> in style attribute then browser will override the red color by blue color and will display the heading in blue color. The same rule is applied for external to internal style definitions.

## 3.14 REVIEW QUESTIONS:-

1. Write short note on HTML. How HTML is different from programming languages?
2. Explain the structure of HTML document.
3. Define tags and attributes.
4. Explain various tags related to text formatting.
5. Explain <OL>,<UL> and <DL> with attributes.
6. Explain <IMG> tag with attributes.
7. Explain various tags for table.
8. Explain hyper links in HTML using <A>.
9. Explain types of hyperlinks using <A>.
10. Explain frames in HTML. Explain <FRAMESET> and <FRAME>.
11. Explain <FORM> with attributes.
12. Explain how <INPUT> tag is used to create different controls.
13. Explain <SELECT> and <TEXTAREA> tags.
14. Explain various attributes of <BODY> tag.
15. Explain the use of <SCRIPT> and <STYLE> tags.

16. Explain following attributes.
    a. Align attribute of <P>
    b. Type attribute of <OL> and <UL>
    c. Src attribute of <IMG>
    d. Cellspacing attribute of <TABLE>
    e. Cellpadding attribute of <TABLE>
    f. Rowspan attribute of <TD>
    g. Colspan attribute of <TD>
    h. Href attribute of <A>
    i. Target attribute of <A>
    j. Cols attribute of <FRAMESET>
    k. Method attribute of <FORM>
    l. Action attribute of <FORM>
    m. Type attribute of <INPUT>
    n. Multiple attribute of <SELECT>
    o. Size attribute of <SELECT>
    p. Behavior attribute of <MARQUEE>
    q. Direction attribute of <MARQUEE>
    r. Scrollamount attribute of <MARQUEE>
    s. Scrolldelay attribute of <MARQUEE>

17. What are style sheets? What are the advantages of using style sheets?
18. Explain types of style sheets.
19. Explain various measurement units in style sheets.
20. Explain how to represent color and URL information in style sheet
21. Explain various text and properties.
22. Explain various margin and padding properties.
23. Explain various background and border properties.
24. Explain various positioning of elements using Position property.

# Chapter 4:- Components Of Dynamic Web

## 4.1 VBScript

- VBScript is developed by Microsoft. It is a light weighted scripting language, a subset of Visual Basic.
- We can write code of VBScript in a text editor without any specific development environment. But because of the interpretation it is comparatively slow.
- Mostly VBScript is used as a server side scripting language in ASP technology. It can also be used as a client side scripting language for Internet Explorer Applications. Some browsers (not developed by Microsoft) either do not support client side scripts or do not support VBScript as a client side scripting language.
- VBScript is a case insensitive language. VBScript supports interpretation. Extra White spaces are automatically ignored by the interpreter.


- <u>Syntax for the Client Side Scripts:-</u>


<SCRIPT> tag inside the <HEAD> section is used to insert client side script. It is compulsory to close the <SCRIPT> tag using </SCRIPT>.

<HTML>

<HEAD>

<SCRIPT LANGUAGE="VBSCRIPT">

…………..

</SCRIPT>

</HEAD>

<BODY>

…………..

</BODY>

- <u>Syntax for the Server Side Scripts:-</u>


Server side script is a part of <BODY> section. It is also possible to write a client side script in <HEAD> section at the same time. We use <% for the beginning of the server side script and %> for the ending of the server side script.

In this section, we will study the basis of VBScript. We will discuss how to use VBScript to write client side and server side scripts in 4.2 and 4.3

### 4.1.1 Data Types and variables:-

VBScript is a weakly typed language and it has only one data type which is Variant (16-bytes). A variant data type can represent empty (declared but uninitialized), byte, integer, long, float, string, date, Boolean, nothing (invalid reference), object etc. values. The main disadvantage of variant data type is the wastage of memory space.

It is not necessary to declare a variable before we use it.VBScript interpreter, by default assumes that any set of characters that is not a keyword is a variable. So if we mistype a variable name then it can cause a problem. If we use Option Explicit statement, then we have to declare a variable using Dim, Public or Private Statements. In this case VBScript interpreter will give error to any unknown sequence of characters.

Dim name

name="Hardik Molia"

Dim birthday

birthday =#1-1-95#

it is required to enclosed the date between #.

Though VBScript has only variant data type, it can hold any of the three kinds of values. Scalar, arrays and object references.

Scalar values are the simple values(integer,string etc.)

### Arrays:-

We can create arrays by using following statements.

| | |
|---|---|
| Dim a1(5) | Declares an array a1 of 5 elements. |
| a2 = Array(10,20,30) | Declares a2 of size 3 and initialize it to 10,20,30 |
| a2(1) | Access to a2(1) which is 20 |
| ReDim a2(25) | ReDim is used to resize an already declared array. Now array a2 is resized to 25 elements from 3 elements. It will erase all the old values. |
| ReDim Preserve a2(35) | Preserves old 25 values |
| Clear array | clears all elements of an array |
| Erase array | removes array from the memory |

74

LBound(array)           Returns the lower bound(index) of an array

UBound(array)           Returns the upper bound(index) of an array

## 4.1.2 Operators:-

VBScript supports following operators.

Arithmetic Operators:-

| Operator | Purpose |
|----------|---------|
| + | Addition |
| - | Subtraction |
| * | Multiplication |
| / | Integer Division |
| \ | Floating Division |
| ^ | Exponential |
| Mod | Modulus |

Relational Operators:-

| Operator | Purpose |
|----------|---------|
| < | Less than |
| > | Greater than |
| <= | Less than or equal to |
| >= | Greater than or equal to |
| = | Equal to |
| <> | Not equal to |

Logical Operators:-

| Operator | Purpose |
|----------|---------|
| And | Logical And |
| Or | Logical Or |
| Not | Logical Not |

String Concatenation:-

| Operator | Purpose |
|----------|---------|
| & | Concatenation of two strings |

## 4.1.3 Inbuilt functions:-

Arguments inside [] are optional.

Mathematical functions:-

| Function Prototype | Example | Result |
|--------------------|---------|--------|
| Abs(number) | Abs(-34.6) | 34.6 |
| Asc(character) | Asc("A") | 65 |
| Eval(expression) | Eval("12+5") | 17 |
| Hex(no) | Hex(10) | "A" |
| Oct(no) | Oct(34) | 42 |
| Fix(no) | Fix(12.34) | 12 |
| Sgn(no) | Sgn(-123) | -1 |
| Sqr(no) | Sqr(4) gives square root | 2 |
| Round(no,[digit]) | Round(123.45565,2) | 123.46 |
| FormatCurrency(no) | FormatCurrency(300) | $300.00 |
| FormatNumber(no,0s) | FormatNumber(12234,3) | 12,234.000 |
| FormatPercent(no) | FormatPercent(0.45) | 45.00% |

Conversion functions:-

| Function Prototype | Example | Result |
|--------------------|---------|--------|
| CBool(expression) | CBool(5>2) | True |
| CByte(number) | CByte(123.45656) | 123 |
| CDate(string) | CDate("August 4, 2011") | 8/4/2011 |
| Chr(ascii) | Chr(65) | "A" |

| CInt(number) | CInt(123.45656) | 123 |
| CStr(number) | CStr(3453) | "3453" |

## Type Checking Functions:-

IsArray(variable),      IsDate(variable),      IsEmpty(variable),

IsNull(variable),      IsNumeric(variable),      IsObject(variable).

All these functions return True if the given variable/value is an array, date, empty, null, number or an object otherwise return False.

## Date/Time Functions:-

Intervals can be specified using yyyy(year), m(month), d(day), w(week), h(hour), n(minute), s(second)

| Function Prototype | Example | Result |
|---|---|---|
| Now | Now | 4/4/2011 9:35:11 PM |
| Time | Time | 10:30:20 PM |
| Date | Date | 4/4/2011 |
| IsDate(date) | IsDate("4/4/2011") | True |
| DateAdd(Interval,Number,Date) | DateAdd("m",1,"1-Jan-11") | 2/1/2011 |
| DateDiff(Interval,Date1,Date2) | DateDiff("m","1-Jan-11","1-Aug-11") | 7 |
| DatePart(Interval,Date) | DatePart("m","1-Jan-11") | 1 |
| DateSerial(Year,Month,Day) | DateSerial(2010,34,4)<br><br>Adds 34 months and 4 days to the year 2010 | 10/4/2012 |
| DateValue(string) | DateValue("May 11, 2011") | 5/11/2011 |
| Hour(Time),Minute(Time) etc. | Hour(Now) | 21 |
| Month(Date),Year(Date) etc. | Year(Date) | 2011 |

| MonthName(no,boolean) | MonthName(10,true) | Oct |
| | MonthName(10,false) | October |
| FormatDateTime(Date,vbgeneral date) | FormatDateTime(Date,vbgeneral date) | 4/4/2011 |
| FormatDateTime(Date,vblongdate ) | FormatDateTime(Date,vblongdate ) | Monday, April 4, 2009 |
| FormatDateTime(Date,vbshortdat e) | FormatDateTime(Date,vbshortdat e) | 4/4/2011 |
| FormatDateTime(now,vblongtime ) | FormatDateTime(now,vblongtime ) | 9:35:52 AM |
| FormatDateTime(now,vbshorttim e) | FormatDateTime(now,vbshorttim e) | 09:35 |

## String Functions:-

| Function Prototype | Example | Result |
| --- | --- | --- |
| LCase(string) | LCase("Hardik") | "hardik" |
| UCase(string) | UCase("Hardik") | "HARDIK" |
| Len(string) | Len("Hardik") | 6 |
| Left(string,no) | Left("Hardik",4) | "Hard" |
| Right(string,no) | Right("Hardik",4) | "rdik" |
| LTrim(string),RTrim(string),Trim(string) | Trim("      Hardik ") | "Hardik" |
| Mid(string,start,[length]) | Mid("VB Script is ", 4, 6) | "Script" |
| Replace(string,find,replace) | Replace("Hi dil","dil","pil") | "Hi pil" |
| Space(no of spaces) | "Hi" & Space(3) "Hardik" | "Hi   Hardik" |
| String(no,character) | String(5,"A") | "AAAAA" |
| StrReverse(string) | StrReverse("Hardik") | "KidraH" |

String Comparison:-

StrComp(string1,string2,option)
Returns 0 if string1 and string2 are same.
Returns 1 if string1 is larger than string2.
Returns -1 if string2 is larger than string2.
For case sensitive comparison, option= vbBinaryCompare

For case insensitive comparison, option= vbTextCompare

Array to String:-

Join(String Array,[delimiter])

This function is used to concate all the elements of an array with a special symbol known as delimiter. It also works with an integer array.

Dim MyString

Dim MyArray(3)

MyArray(0) = "Hardik "

MyArray(1) = "Molia "

MyArray(2) = "Works At "

MyArray(3) = "GEC Rajkot"

"Hardik Molia Works At GEC Rajkot" = Join(MyArray)

"Hardik *Molia *Works At *GEC Rajkot = Join(MyArray,"*")

String to Array:-

Split(string,splitting character)
This function is used to split a string in to a set of elements of an array. The argument is the string having words separated by a splitting character.

Dim MyString, MyArray, Msg
MyString = "VBScriptXisXfun!"
MyArray = Split(MyString, "x")
MyArray(0) will contain "VBScript".
MyArray(1) will contain "is".
MyArray(2) will contain "fun!".


Search in a String:-

InStr(String,search)

InStr(Startindex,String,search,mode)

This function returns the index of the first match of search text in string. Index starts with 1. By defualt starting index is 1. Mode is optional Textcompare or binarycompare. InStr starts searching from the beginning of the string(left to right)

InStrRev(String,search)

InStrRev(String,search, Startindex,mode)

InStrRev starts searching from the end of the string(right to left)

name1 = "this is the world of computer. it is very useful"

| | |
|---|---|
| instr(name1,"is") | returns 3 |
| instr(4,name1,"is") | returns 6 |
| instrrev(name1,"is",7) | returns 6 |
| instr(1,name1,"IS",vbtextcompare) | returns 3 |

### 4.1.4 Constants:-

There are some default constants which we can use to avoid the use of numerical values. Here are some of the constants which we use frequently.

Color constants:-vbBlack, vbRed, vbGreen etc.

Comparision constants:-vbBinaryCompare, vbTextCompare

Date/Time Constants:-vbSunday,vbMonday etc.

We can declare a user defined constant using const keyword

Const pi = 3.14

Const subject="Dynamic Web Development"

### 4.1.5 Branching Statements:-

If.... Then....

There are many variations of if... then... are possible. Here are some of them with examples.

| If I mod 2 = 0 Then | If I > 0 Then |
|---|---|
| Response.Write "I is Even" | Response.Write "I is positive number" |
| Else | ElseIf I < 0 Then |

| Response.Write "I is Odd"<br><br>End If | Response.Write "I is negative number"<br><br>Else<br><br>Response.Write "I is zero"<br><br>End If |
|---|---|

Select... Case....

If we want a statement which selects a matching case from many possible cases then it is better to use Select...Case... statement instead of nesting of If...Then...

```
Select Case branchcode

Case 11

        Response.Write "I.T."

Case 12

        Response.Write "E.C."

Case Else

        Response.Write "Unknown Code"

End Select
```

## 4.1.6 Looping Statements:-

There are basically five looping statements which provide iterative execution of the statements. Here is an example to calculate sum of 1 to 10 using different looping statements.

| Sum=0<br><br>I=1<br><br>Do While I<=10<br><br>Sum = Sum + I<br><br>I = I + 1<br><br>Loop | Sum=0<br><br>I=1<br><br>Do Until I=10<br><br>Sum = Sum + I<br><br>I = I + 1<br><br>Loop |
|---|---|

| | |
|---|---|
| Sum=0<br><br>I=1<br><br>While I<=10<br><br>Sum = Sum + I<br><br>I = I + 1<br><br>Wend | Sum=0<br><br>For I = 1 to 10 Step 1<br><br>Sum = Sum + I<br><br>Next<br><br><br>'Step 1 indicates increment by 1(Default). Similar to I++ in C. we can specify any positive or negative number. |

We often need to use a loop for an array. One possible way is to use UBound(array) to find out the size of an array. One another way is to use for each... loop.

Assume that array mydata(5) contains some numbers. We want to sum all those numbers.

| | |
|---|---|
| Sum=0<br><br>For I = 1 to UBound(mydata) Step 1<br><br>Sum = Sum + mydata(I)<br><br>Next | Sum=0<br><br>For each I in mydata<br><br>Sum = Sum + I<br><br>Next |

It is also possible to avoid writing same object name more than once by using With... End With...

| | |
|---|---|
| Response.Write "Hello<BR>"<br><br>Response.Write "How Are You"<br><br>Response.Write "Good Bye" | With Response<br><br>.Write "Hello<BR>"<br><br>.Write "How Are You"<br><br>.Write "Good Bye"<br><br>End With |

## 4.1.7 Function & Subroutine:-

We can create a functio(returns a value) and a subroutine(does not return a value) in VBScript. The function returns a value stored in the variable name same to the function name. We use Function keyword for a function and Sub keyword for a subroutine. Here is an example of finding maximum out of two numbers.

| Function max(no1,no2) | Sub max(no1,no2) |
|---|---|
| If no1>no2 Then | If no1>no2 Then |
| max = no1 | Response.Write no1 |
| Else | Else |
| max = no2 | Response.Write no21 |
| End If | End If |
| End Function | End Sub |
| Calling statement:- | Calling statement:- |
| Ans=max(5,7) | max(5,7) or max 5,7 |

## 4.1.8 InputBox and MsgBox:-

These two functions are used interact with the user through alert boxes. We can use these statements in client side scripts only.

**InputBox:-**InputBox is used to get information from the user. User enters the information in a text box provided by the InputBox.

InputBox(prompt[,title][,default][,xpos][,ypos])  Arguments  inside  []  are optional.

| Parameter | Description |
|---|---|
| prompt | The message to show in the dialog box. |
| title | The title of the dialog box |
| default | A default text in the text box of input box. |
| xpos | The prompt box' distance from the left edge of the screen. |
| ypos | The prompt box' distance from the top edge of the screen. |

**MsgBox:-**

MsgBox(prompt[,buttons][,title]) Arguments inside [] are optional.

| Parameter | Description |
|-----------|-------------|
| prompt | The message to show in the dialog box. |
| Buttons | It specifies number and types of buttons to display. |
| title | The title of the dialog box |

Buttons argument represents the types of buttons and an icon for the message box. This argument can be represented by using following constants.

Constants for message box style:-

vbOkCancel          vbAbortRetryCancel               vbYesNoCancel       vbYesNo

vbRetryCancel       vbOkOnly

Constants for message box icon:-

vbQuestion          vbExclamation            vbInformation              vbCritical

Constants for buttons:-

vbOk -          1

vbCancel -      2

vbAbort -       3

vbRetry -       4

vbIgnore -      5

vbYes -         6

vbNo -          7

| Value of Buttons | Description |
|------------------|-------------|
| vbOkOnly+ vbInformation | OK Button with an icon of Information |
| vbOkCancel + vbCritical | OK and CANCEL buttons with an icon of critical. |

This function is used as a function as well as a subroutine.

## MsgBox as a subroutine:-

User will get a simple message box with a message. When user will click on any of the given buttons, browser will resume the execution of the client side script. It is not possible to identify the user has clicked which button.

Use:- Just to show some information with Ok button only.

## MsgBox as a function:-

User will get a simple message box with a message. When user will click on any of the given buttons, program will get the value corresponding to the button clicked by the user.

Use:- Get answer in Yes/No from the user.


## 4.2 SERVER SIDE AND CLIENT SIDE SCRIPTING:-

In real life applications, we should decide whether to use client side script or server side script according to the purpose of using it.

Suppose we want to design a web form for new user account registration. So we will design a web form with HTML tags having some fields (name, surname, address, password email address etc.). The basic requirement of creating a new account is that all fields must be properly filled (all fields must have valid values entered by the user). If all fields are properly filled by the user, then only it should be possible to create a new account, otherwise it should provide some error messages to the user.

The above requirement can be implemented using both the types of scripts (client side, server side)

Solution using Server Side Script:-


1. Create a web form with required fields using <HTML> tags.
2. Create an asp file, contain some code in VBScript to verify whether all field values entered by the user are valid or not. If all are valid then do the registration otherwise provide which values are invalid. (We will discuss how to write such code in chapter 5.)
3. Server side scripts are executed by the web server and so the main problem with the above solution is that it increases the execution load on the server.
4. Whenever a new user will try to do the registration, first of all, server will provide him the form. Then whenever he will try to submit the form, server will get all the values entered by the user. Server will verify it by executing the server side script. Now if all the fields are valid then script will create a new account. But if all the fields are not valid then script will

generate a reply with some error messages for invalid fields. Server will send this reply back to the client and the user will able to see which fields were invalid in his request. Now again server will repeat this process if user will try to submit the form without entering valid values for all the fields.

5. So this solution increase the client-server interaction(sending of error messages from server to client and sending of form values from client to each time a user enters an invalid value) as well as the burden of execution at server(each time server will get the form values, it will execute a server side script to verify them)

6. So this solution is not the best way to use specially when the total number of simultaneous users who are accessing your web site is large.

Solution using Client Side Script:-

1. The main purpose of using client side scripts is to give certain responsibility to the client browser before it sends a new request to the server.

2. Create a web form with required fields using <HTML> tags.

3. Instead of writing scripting code to verify the validity of the values entered by the users in another file as a server side script, include the code in the file of web form as a client side script.

4. Create a page which contains a script to accept values from the user and creates the user account.

5. So now the server will only execute the script which creates a user account.

6. Client side code of form verification will be executed by the client browser. We should write such a script in a way that when user will submit the form, client browser will execute the client script. The script will check the validity of all the form fields. If all fields are valid then browser will send the values to the server for further processing(in our example, execution of a script which creates the user account) otherwise browser will execute some code which informs user about the invalid fields (By using message box) and will stop further submission of form to the server.

7. So now we have assigned the responsibility of verifying the validity of all the form fields to the client browser. So it reduces both client-server interaction and server side execution.

Let's discuss another example:-

Suppose in above example, one more important feature is required. The username must be valid (No spaces, Minimum length is 5 characters etc.) and must be available (Not already registered by some other user).

Suppose we have a database on the server which keeps record about the username and passwords. In our example we should verify the username, entered by the user for its validity as well as for its availability.

Username's validity can be verified by a client side script because it is independent of all the already registered usernames.

Username's availability can not be verified by a client side script because the database is on the server. And so we can verify the availability by using a server side script.

## 4.3 SERVER SIDE SCRIPT EXAMPLES:-

We will see how to read values from the user in chapter 4. in this section, we will study some simple examples based on server side scripts.

4.1.asp (Percentage and Grade Calculation)

| <HTML>\n<body>\n<%\nDim marks1,marks2,marks3\nDim percentage\nmarks1=56\nmarks2=76\nmarks3=55\npercentage=(marks1+marks2+marks3)/3\nResponse.Write "Percentage:- " &\nRound(percentage,2) & "%"\nResponse.Write "<BR>"\n\nIf percentage>=70 Then\nResponse.Write "Distinction"\nElseIf percentage>=60 Then\nResponse.Write "First Class"\nElseIf percentage>=50 Then\nResponse.Write "Second Class"\nElseIf percentage>35 Then\nResponse.Write "Third Class"\nElse\nResponse.Write "Fail"\nEnd If\n%>\n</BODY>\n</HTML> | OUTPUT:-<br><br>Percentage:- 62.33%<br>First Class |
|---|---|

## 4.2.asp (Multiplication Table)

| | OUTPUT:- |
|---|---|
| ```<br><HTML><br><BODY><br><%<br>dim no,i<br>no = 10<br>Response.Write "<TABLE>"<br>For i = 1 To 10<br>Response.Write "<TR>"<br>Response.Write "<TD>" & no & " <TD>*<TD>" & i<br>Response.Write  "<TD>=<TD>" & no*i & "<TR>"<br>Next<br>Response.Write "</TABLE>"<br>%><br></BODY><br></HTML><br>``` | $10 * 1 = 10$<br>$10 * 2 = 20$<br>$10 * 3 = 30$<br>$10 * 4 = 40$<br>$10 * 5 = 50$<br>$10 * 6 = 60$<br>$10 * 7 = 70$<br>$10 * 8 = 80$<br>$10 * 9 = 90$<br>$10 * 10 = 100$ |

## 4.3.asp (Fibonacci Series)

| | OUTPUT:- |
|---|---|
| ```<br><HTML><br><BODY><br><%<br>Dim curval,prevval,newval<br>curval=1<br>preval=0<br>newval=1<br>Response.Write "<H3>"<br>Response.Write preval & "<BR>" & curval<br><br>For i = 1 To 10<br>        curval=preval<br>        preval=newval<br>        newval=curval+preval<br>        Response.Write "<BR>" & newval<br>Next<br>%><br></BODY><br></HTML><br>``` | 0<br>1<br>1<br>2<br>3<br>5<br>8<br>13<br>21<br>34<br>55<br>89 |

## 4.4 CLIENT SIDE SCRIPT EXAMPLES:-

In client side scripts we use Document object instead of Response object.

4.4.asp (Percentage and Grade Calculation)

| | OUTPUT:- |
|---|---|
| <pre><HTML><br><HEAD><br><SCRIPT LANGUAGE=VBSCRIPT><br>Dim marks1,marks2,marks3<br>Dim percentage<br>marks1 = CInt(InputBox("Enter marks of DWD"))<br>marks2 = CInt(InputBox("Enter marks of WC"))<br>marks3 = CInt(InputBox("Enter marks of JAVA"))<br>percentage=(marks1+marks2+marks3)/3<br>Document.Write "Percentage:- " & Round(percentage,2)<br>& "%"<br>Document.Write "<BR>"<br><br>If percentage>=70 Then<br>Document.Write "Distinction"<br>ElseIf percentage>=60 Then<br>Document.Write "First Class"<br>ElseIf percentage>=50 Then<br>Document.Write "Second Class"<br>ElseIf percentage>35 Then<br>Document.Write "Third Class"<br>Else<br>Document.Write "Fail"<br>End If<br><br></SCRIPT><br></HEAD><br></BODY><br></HTML></pre> | Percentage:- 62.33%<br>First Class |

## 4.5.asp (Multiplication Table)

| | OUTPUT:- |
|---|---|
| ```html<br><HTML><br><HEAD><br><SCRIPT LANGUAGE=VBSCRIPT><br>dim no,i,lowerlimit,upperlimit<br>no = InputBox("Enter number")<br>lowerlimit = InputBox("Enter lower limit")<br>upperlimit = InputBox("Enter upper limit")<br>Document.Write "<TABLE>"<br><br>For i = lowerlimit To upperlimit<br>        Document.Write "<TR>"<br>        Document.Write "<TD>" & no & " <TD>*<TD>"<br>& i & "<TD>=<TD>" & no*i & "<TR>"<br>Next<br>Document.Write "</TABLE>"<br></SCRIPT><br></HEAD><br></BODY><br></HTML><br>``` | 5 * 7  = 35<br><br>5 * 8  = 40<br><br>5 * 9  = 45<br><br>5 * 10 = 50<br><br>5 * 11 = 55<br><br>5 * 12 = 60<br><br>5 * 13 = 65<br><br>5 * 14 = 70<br><br>5 * 15 = 75<br><br>5 * 16 = 80 |

## 4.6.asp (Fibonacci Series)

| | OUTPUT:- |
|---|---|
| ```html<br><HTML><br><HEAD><br><SCRIPT LANGUAGE=VBSCRIPT><br>dim curval,prevval,newval,terms<br>terms=InputBox("Enter total terms")<br>curval=1<br>preval=0<br>newval=1<br>Document.Write "<H3>"<br>Document.Write preval & "<BR>" & curval<br>For i = 3 To terms<br>        curval=preval<br>        preval=newval<br>        newval=curval+preval<br>        Document.Write "<BR>" & newval<br>Next<br></SCRIPT><br></HEAD><br></BODY><br></HTML><br>``` | 0<br>1<br>1<br>2<br>3<br>5<br>8<br>13<br>21<br>34<br>55<br>89 |

90

## 4.7.asp (Calculate age from the birth date and print appropriate message)

| | OUTPUT:- |
|---|---|
| ```<HTML>``` | |
| ```<HEAD>``` | |
| ```<SCRIPT LANGUAGE=VBSCRIPT>``` | |
| ```Dim bday,diff``` | Your Age is - 21 |
| ```bdate = InputBox("Birth Date...")``` | You are a youngster... |
| ```If IsDate(bdate) Then``` | |
| ```diff = (DateDiff("m",CDate(bdate),Date))\12``` | |
| ```Document.Write "Your Age is - " & diff & "<BR>"``` | |
| | |
| ```If dIff <=5 Then``` | |
| ```    Document.Write "You are a kid..."``` | |
| ```ElseIf dIff <=18 Then``` | |
| ```    Document.Write "You are a school boy..."``` | |
| ```ElseIf dIff <=25 Then``` | |
| ```    Document.Write "You are a youngster..."``` | |
| ```ElseIf dIff <=40 Then``` | |
| ```    Document.Write "You might be a married person..."``` | |
| ```Else``` | |
| ```    Document.Write "You are an old man..."``` | |
| ```end If``` | |
| ```end If``` | |
| ```</SCRIPT>``` | |
| ```</HEAD>``` | |
| ```</BODY>``` | |
| ```</HTML>``` | |

## 4.8.asp (Sorting of numbers in descending order)

| | OUTPUT:- |
|---|---|
| ```<HTML>``` | |
| ```<HEAD>``` | |
| ```<SCRIPT LANGUAGE=VBSCRIPT>``` | |
| ```Dim total,i,j,temp``` | 87 |
| ```Dim vals(5)``` | 76 |
| ```total = 5``` | 56 |
| ```For i = 0 To total-1``` | 45 |
| ```    vals(i) = InputBox("Input Number.")``` | 34 |
| ```next``` | |
| | |
| ```For i = 0 To total-1``` | |
| ```    For j = 0 To total - 1``` | |
| ```        If vals(i) > vals(j) Then``` | |
| ```            temp = vals(i)``` | |

```
                    vals(i) = vals(j)
                    vals(j) = temp
             End If
        Next
Next

For i = 0 To total-1
      Document.Write vals(i) & "<BR>"
Next
</SCRIPT>
</HEAD>
</BODY>
</HTML>
```

4.9.asp (Square roots of 1 to N)

```
<HTML>
<HEAD>
<SCRIPT LANGUAGE=VBSCRIPT>
dim i,terms
terms=InputBox("Enter total terms")
Document.Write "<TABLE
BORDER=1><TR><TD>NO<TD>SQUARE ROOT"

For i = 1 To terms
      Document.Write "<TR>"
      Document.Write "<TD>" & i & "<TD>" & Sqr(i)
Next
      Document.Write "</TABLE>"

</SCRIPT>
</HEAD>
</BODY>
</HTML>
```

**OUTPUT:-**

| NO | SQUARE ROOT |
|----|-------------|
| 1 | 1 |
| 2 | 1.4142135623731 |
| 3 | 1.73205080756888 |
| 4 | 2 |
| 5 | 2.23606797749979 |
| 6 | 2.44948974278318 |
| 7 | 2.64575131106459 |
| 8 | 2.82842712474619 |
| 9 | 3 |
| 10 | 3.16227766016838 |

## 4.10.asp (Factorial Series 1! 2! 3! 4!.....N!)

| | OUTPUT:- |
|---|---|
| ```<HTML>``` <br> ```<HTML>``` <br> ```<HEAD>``` <br> ```<SCRIPT LANGUAGE=VBSCRIPT>``` <br> ```dim total,i,k``` <br> ```i=65``` <br><br> ```total = InputBox("Input total terms...")``` <br> ```For k = 1 To total``` <br> ```        Document.Write cal(k) & "<BR>"``` <br> ```Next``` <br><br> ```Function cal(no)``` <br> ```Dim i``` <br> ```cal=1``` <br> ```For i= no To 1 Step -1``` <br> ```cal = cal * i``` <br> ```Next``` <br> ```End Function``` <br><br> ```</SCRIPT>``` <br> ```</HEAD>``` <br> ```</BODY>``` <br> ```</HTML>``` | 1 <br> 2 <br> 6 <br> 24 <br> 120 <br> 720 |

## 4.11.asp (Search total number of occurrences of a string)

```
<HTML>
<HEAD>
<SCRIPT LANGUAGE=VBSCRIPT>

Dim total,i,j,temp,ch,index

inp = InputBox("Enter String...")
ch = InputBox("Enter Character..")

total = 0
For i = 1 To Len(inp)
      index =  instr(i,inp,ch)
      If index <> 0 Then
      i = index + 1
      total = total + 1
      End If
```

```
Next

Document.Write "<BR>Message String:- " & inp
Document.Write "<BR>Search  String:- " & ch
Document.Write "<BR>Total Occurrences:- " & total

</SCRIPT>
</HEAD>
</BODY>
</HTML>
```

**OUTPUT:-**

Message String :- This is the world of computer.
Search String :- is
Total Occurrences:- 2

4.12.asp (Change the case of a text to Title Case(first letter of each word in capital)

```
<HTML>
<HEAD>
<SCRIPT LANGUAGE=VBSCRIPT>
dim str,i,res,first, second
str = Inputbox("Enter Sentence...")
Dim parts
parts = split(str," ")
res = ""

For Each part In parts
        first = ucase(left(part,1))
        second = lcase(mid(part,2))
        res = res & first & second & " "
Next

MsgBox res,vbOkOnly+vbInformation,"Title Case"

</SCRIPT>
</HEAD>
</BODY>
</HTML>
```

**OUTPUT:-**

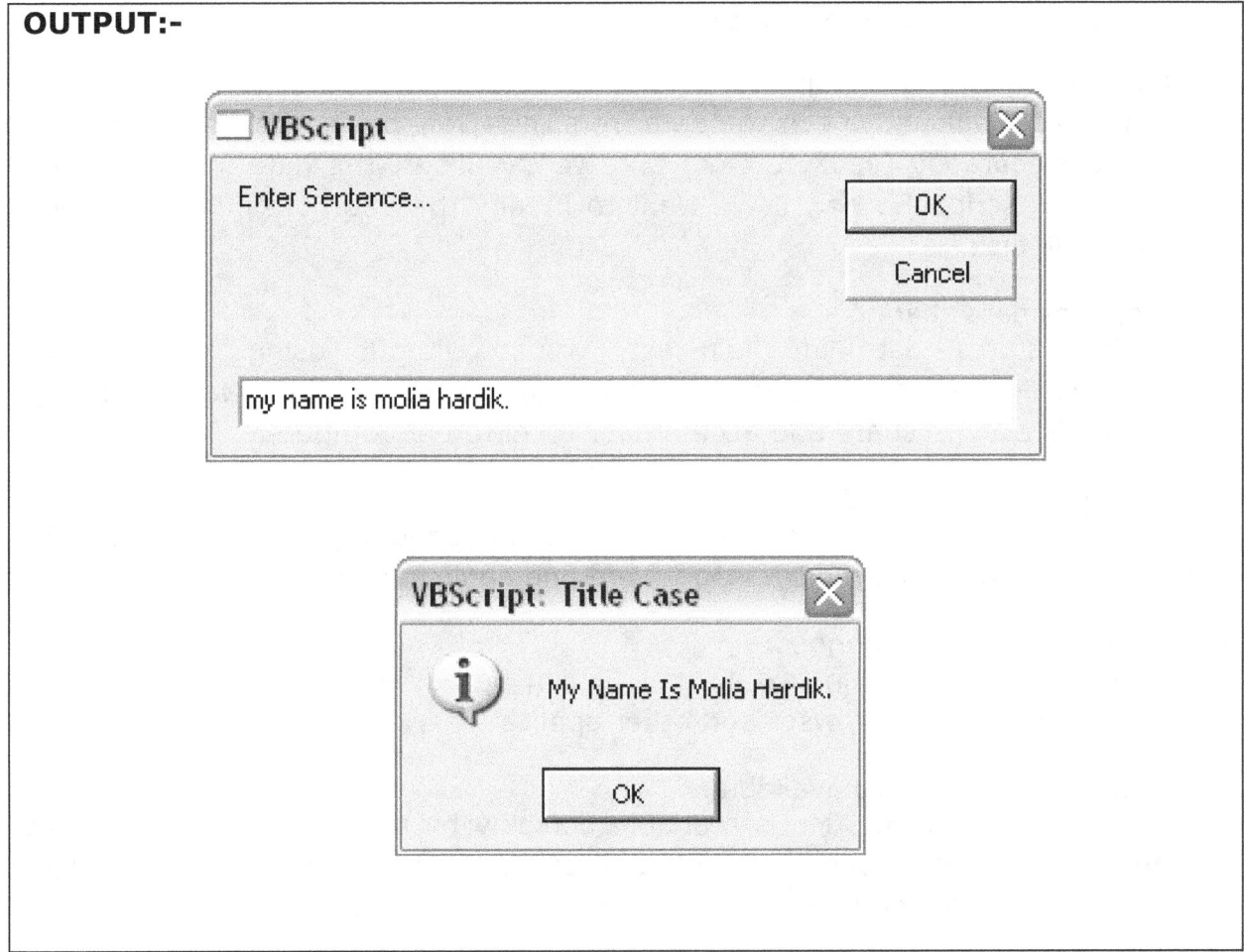

## 4.5 XML:-

**XML stands for eXtensible Markup Language developed by World Wide Web Consortium.**

| HTML | XML |
|---|---|
| HTML stands for Hyper Text Markup Language | XML stands for eXtensible Markup Language. |
| HTML is used to display data | XML is used to carry data |
| The main focus is how data looks. | The main focus is what data is. |
| HTML tags are predefined. We can not create our own tags. | XML has no predefined tags. We create our own tags. |
| HTML can not be used to describe other languages. | XML can be used to describe other languages. |
| HTML is case insensitive. | XML is case sensitive. |
| HTML allows omitting the closing tags. | XML does not allow to omit the closing tags. |
| There is no need to proper nesting of tags.<B><I><U>Hello</I></U></B> is valid | All tags must be properly nested. <B><I><U>Hello</I></U></B> is invalid |
| Attribute values required to be placed in quotations only when they contain spaces. | Attribute values always required to be placed in quotations. |

### 4.5.1 Advantages of using XML:-

- Separates data from HTML:-
  It is very difficult to edit the data from the HTML pages. Using XML we can store data separately (in .XML files) and we can insert the data in HTML page using a small script. So now if we want to modify the data, we need to modify only the XML files.

- Simplifies data sharing:-
  XML stores data in plain-text format without using any specific software/hardware. So it is the software/hardware independent way of storing data. So it is easy to share and access data on different computers.

- Simplifies data transport:-
  Because of the plain-text and software independent format of XML, it is very less time consuming to transport from one application to another.

- Simplifies platform changes:-
  Because of the plain text format, it can be easily used by the upgraded operating system, web browser and other applications.

- Makes your data more available:-
  Data stored using XML is readable not only by the web applications, but by the desktop applications also. XML data can easily read by all kind of reading machines (handheld computers, voice machine)

- Creates new languages for web designing:-
  XML is also used to represent other languages for web designing. XHTML(stricter and cleaner version of HTML) ,WSDL(for web services) WAP and WML (for portable mobile devices) etc.

### 4.5.2 XML Syntax:-

- All XML Elements Must Have a Closing Tag.
  <Name> must ends with a corresponding </Name>

- XML Tags are Case Sensitive.
  <Name> and <name> are different.

- XML Elements Must be Properly Nested.
  <Name>Hardik<Surname>Molia</Name></Surname> is invalid

  <Name>Hardik<Surname>Molia</Surname></Name> is valid

  In above example, <Surname> is openend inside <Name> so it must be closed inside <Name>

- XML Documents Must Have a Root Element.
  XML document starts with a root tag as a parent tag which defines all other tags as child tags in a tree structure.

- XML Attribute Values Must be quoted
    <Employee Name=Hardik> is invalid.

    <Employee Name="Hardik"> is valid.

### 4.5.3 XML Tree Structure:-

XML document forms a tree structure which starts with a root (parent) node and ends with leaves (child) nodes.

4.13.xml

```
<?xml version="1.0" encoding="ISO-8859-1"?>
<personalinfo>
    <name>Hardik</name>
    <surname>Molia</surname>
    <city>Rajkot</city>
</personalinfo>
```

<?xml version="1.0" encoding="ISO-8859-1"?> specifies the XML version and encoding scheme.

<personalinfo> is the root element of the document. It has three child elements <name>, <surname> and <city>.

Document ends with the closing root element </personalinfo>

4.14.xml

```
<?xml version="1.0" encoding="ISO-8859-1"?>
<bookstore>
    <book category="programming">
        <title lang="HTML">learn HTML</title>
        <author>xyz</author>
        <year>2008</year>
        <price>300</price>
    </book>
    <book category="programming">
        <title lang="PHP">learn PHP</title>
        <author>abc</author>
        <year>2010</year>
        <price>500</price>
    </book>
    <book category="operating system">
        <title lang="english">OS Concepts</title>
        <author>xyz</author>
        <year>2005</year>
        <price>300</price>
```

```
    </book>
</bookstore>
```

<bookstore> is the root element. It has three child elements (<book> with an attribute category). Each child element has four sub child elements. <title>,<author>,<year> and <price>

### 4.5.4 Viewing XML file:-

XML files are not executable so we can not execute them. We can view the XML file in any browser. The browser displays the xml file with color coded elements. A plus (+) or (-) sign can be used to expand or collapse the child elements.

If we open an erroneous xml file, browser will display an error message.

### 4.6 XSL:-

* XSL stands for eXtensible Stylesheet Language – an XML based style sheet language developed by the World Wide Web Consortium (W3C).

* A XML document doesn't contain any information regarding how browser will display it. XSL describes how the XML document should be displayed.
* XSL consists of three parts:-
  o XSLT - the language for transforming XML documents
  o XPath - the language for navigating in XML documents
  o XSL-FO - the language for formatting XML documents

* We study the XSTL which is used for transforming XML document into XHTML or other XML documents.
* XSLT transforms an XML document into XHTML by transforming each XML element into XHTML element.

### Example 4.14 Transform XML document into XHTML document using XSLT

Step 1:- We want to transform 4.14.xml file into XHTML

4.14.xml

```
<?xml version="1.0" encoding="ISO-8859-1"?>
<?xml-stylesheet type="text/xsl" href="4.14style.xsl"?>
<personalinfo>
    <name>Hardik</name>
    <surname>Molia</surname>
    <city>Rajkot</city>
</personalinfo>
```

- <?xml-stylesheet type="text/xsl" href="4.14style.xsl"?> specifys that we want to display this file with the style definations of 4.14style.xsl

Step 2:- Create an XSL style sheet 4.14style.xsl with transformation template.

4.14style.xsl

```
<?xml version="1.0" encoding="ISO-8859-1"?>
<xsl:stylesheet version="1.0"
mlns:xsl="http://www.w3.org/1999/XSL/Transform">

<xsl:template match="/">

<html>
<body>
<h2>My Personal Information</h2>
<table border="1">

<xsl:for-each select="personalinfo">

<tr><td>Name:-</td><td><xsl:value-of select="name"/></td></tr>
<tr><td>Surname:-</td><td><xsl:value-of select="surname"/></td></tr>
<tr><td>City:-</td><td><xsl:value-of select="city"/></td></tr>

</xsl:for-each>

</table>
</body>
</html>
</xsl:template>
</xsl:stylesheet>
```

- 4.14style.xsl contains the XHTML template. It starts with the information of XML version, XSL version.

- The template information begins with <html> and other HTML tags. To insert data from the XML file, use tags of XHTML.

- <xsl:template match="/"> starts the template with "/" defines the whole document. If we want to use only an element we can specify its name in match attribute.

- <xsl:for-each select="personalinfo"> works as a loop for all the child elements.

- <xsl:value-of select="name"/> gives the value of the child element "name" of "personalinfo".

- We can use the above statements in nesting of loops according to the tree structure of our XML document.

Step 3:- View 4.14.xml in a browser.

---

**OUTPUT:-**

# My Personal Information

| Name:- | Hardik |
|--------|--------|
| Surname:- | Molia |
| City:- | Rajkot |

---

## 4.7 REVIEW QUESTIONS:-

1. Explain variables and data types in VBScript.
2. Explain types of operators and statements in VBScript.
3. Differentiate a function and a subroutine in VBScript.
4. Explain following functions with examples.

    a. Formatcurrency()
    b. Dateadd()
    c. Datediff()
    d. Datepart()
    e. Dateserial()
    f. Monthname()
    g. Formatdatetime()
    h. Trim()

5. Explain InputBox function.
6. Explain MsgBox as a function and as a subroutine.
7. Why we need to use scripting language? Explain the importance of client side and server side scripts.
8. Differentiate client side and server side scripts.
9. Differentiate HTML and XML.
10. Explain advantages of using XML to represent data.
11. Explain tree structure in XML.
12. What is XSL?
13. Explain how to convert an XML document into XHTML using XSLT.

# Chapter 5:- Active Server Pages

## 5.1 INTRODUCTION TO ACTIVE SERVER PAGES.

ASP is a technology developed by Microsoft. It is not a programming language itself.ASP works with scripting languages.

ASP is a server side scripting environment which is used to design dynamic, individualized and content based interactive web sites.

ASP is used to design dynamic and content based web pages. Server will create the web page according to the request of a user. User enters a roll no and server sends a web page containing information about his result.

ASP is used to design web pages which are interactive and individualized. User can set some parameters for text formatting. Server will identify a particular user and will create the web page with the formatting specified by that user.

**Need for ASP:-**

**1) ASP objects:-**

ASP supports various intrinsic built-in objects which are always available to use. We don't need to create them manually. Using ASP objects, we can collect information from the client, send information to the client, collect information about the web server, manage session and application variables etc.

**Example:-**Retrieve information from the client browser which user has entered in various input fields of a form and provide the next page accordingly.

**2) Database Access:-**

ASP supports retrieval of information from various databases (SQL server, Oracle, Access etc.) using ActiveX Data Object (ADO).

**Example:-** Online results, Online ticket booking

**3) Recognizing Individuals:-**

ASP can provides personalized content by differentiating one browser from another. ASP accomplishes this through Session object and cookies. Using these features, ASP identifies a specific user on a specific computer and provides the content accordingly.

**Example:-** Save username and passwords through browser so we don't need to enter every time we want to login. Display the username in all the pages after successful logged in.

## 4) State Maintenance:-

This is the process of keeping track of a user's progress through your site or application.

**Example:-** Keeping track of how many users have visited our web site for how much duration.

## 5) ASP Extensibility:-

Using Server & ContextObject object your application can create and access external COM components.

**Example:-** Use of adrotator control in a web page.

## Advantages of ASP:-

- ASP code resides in text files.
- ASP code times out after 90 sec. (Solution to endless execution) (time can be changed)
- ASP code is server safe (runs in a limited space)
- ASP applications are small can be developed using many scripting languages.
- We can modify ASP applications without stopping the server.

## Disadvantages of ASP:-

- ASP supports interpretation not compilation. So it is slow.
- ASP supports programming using markup languages (HTML) and scripting languages (VB Script) which are small and having less functionality as compared to higher level programming languages.
- All variable are variants so they are larger and slower than typed variables.
- ASP supports late data binding which is not suitable for the processing of large amount of data.

## 5.2 ASP MODEL (DYNAMIC WEB PAGE HANDLING BY WEB SERVER)

In Chapter 1 we discussed about the static and dynamic web page design. In this topic we will see how server will process the requests of web pages.

### Web Request

User can request a web page from server by entering the specific URL in the address bar of the browser through which he wants to see the web page. User can enter the URL manually or by clicking some hyper link.

URL has the domain name of the server. It is required to find out the corresponding ip address of that server. When browser makes a request it sends the URL to a naming server. A naming server finds the IP address (like

203.55.333.45) corresponding to the domain name of the server and relay request to that server. If it doesn't find any proper IP address for a domain name, then it will send an error message to the client browser.

## Server Response

Server receives the request of a page. Server serves requests in first come first served basis.

URL contains the virtual path of a web page. Server will translate the virtual path to the physical path and will find out the address of the requested web page.

Server will check is it possible to serve the request of that page at this moment or not. Because of various reasons, sometimes server cannot process the request and in such cases, it will send an error message with an error code to the client browser which requested for that page. (HTTP 404 error – file not found or HTTP 403 error for access denied etc.)

The requested page can be of any extension. Some requires execution while some do not. If a user requests for an mp3 file or a jpg file, then these requests are the static file requests. If a user requests for a HTML document then it is also a static web page request. But if a user requests for an ASP page then it is a dynamic web page request.

Server will identify the type of the request. If it is static then server will not need to execute it. Server will send that file back to the client.

If the request is for the dynamic web page, then server will execute it, the result of the execution will be a page in HTML. Server will send this resultant HTML page back to the client. This is the response of the server to the request of the client. During the execution server may need to access database, some other files, and some other components as per the code written in that page.

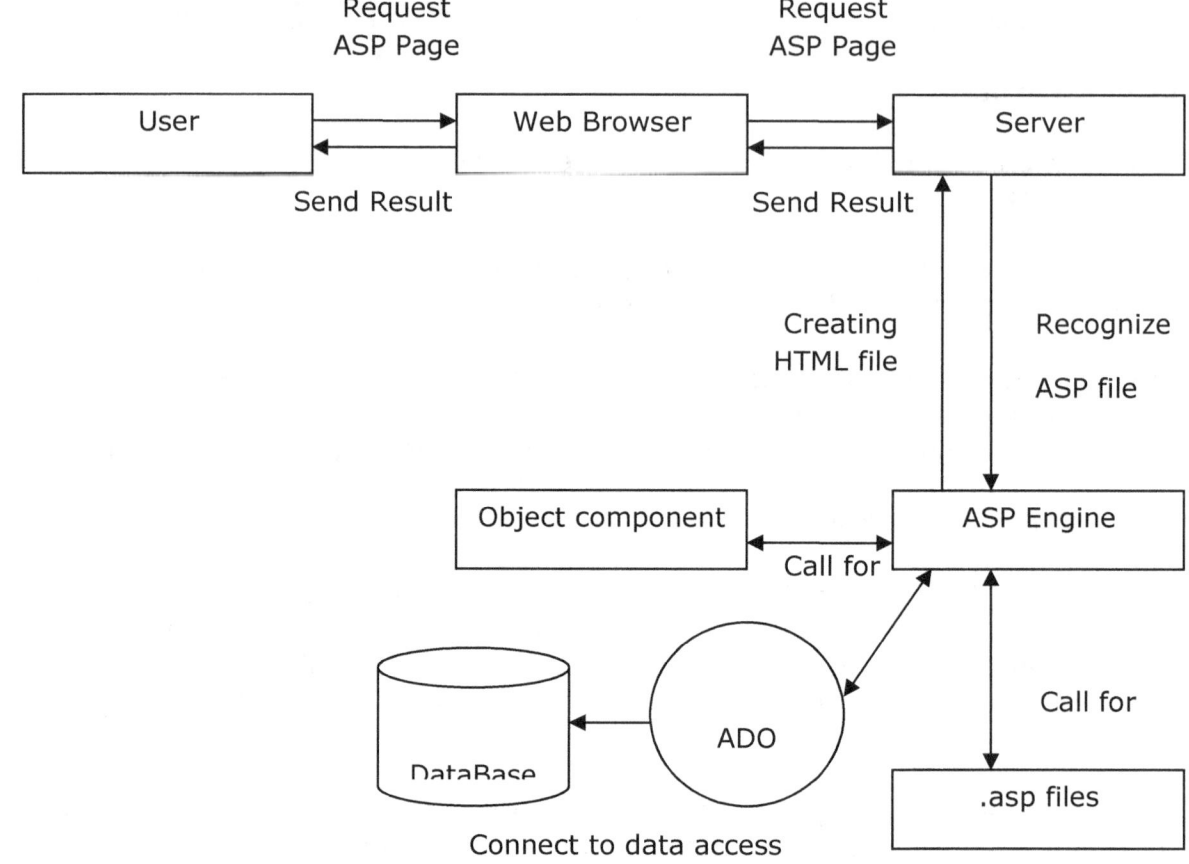

## 5.3 ASP COMPATIBILITY

**Web server:-**A web server is a combination of hardware and software which provides various facilities to deliver content that we can access through the Internet.

**Personal Web Server (PWS):-**PWS is a software, developed for the windows operating systems to provide an environment for the execution of content based on ASP technology. The PWS is compatible with Windows 95, Windows 98 and Windows NT (prior to the windows 2000). PWS supports various protocols like HTTP, SMTP, FTP.

**Internet Information Services (IIS):-**IIS is the replacement of the PWS from windows 2000 onwards. Microsoft has released a new modified version of IIS with a new version of Microsoft's windows operating systems. The latest version of IIS is the IIS 7.5 which is an integral part of Windows 7.it supports HTTP, HTTPS, FTP, FTPS, SMTP, NNTP. From windows vista onwards, IIS is an integral part of the windows operating system. In older operating systems (Windows XP), we can install the IIS from the Source CD of operating system.

## 5.4 RUN ASP ON PC

To run ASP programs in our system, we need our computer to work as a client machine as well as a web server. To make it work as a web server, we need to have IIS.

### Install IIS on Windows Vista/Windows 7

1. Open the Control Panel
2. Open Programs and Features
3. Click "Turn Windows features on or off"
4. Select the check box for Internet Information Services (IIS), and click OK
5. Open the control Panel
6. Open the Administrative Tools
7. Check the shortcut of IIS.

### Install IIS on Windows XP /Windows 2000

1. Open the Control Panel
2. Open Add or Remove Programs
3. Click Add/Remove Windows Components
4. Click Internet Information Services (IIS)
5. Click Details
6. Select the check box for World Wide Web Service, and click OK
7. In Windows Component selection, click Next to install IIS
8. You may need to insert the source CD of windows XP/2000.
9. Open the control Panel
10. Open the Administrative Tools
11. Check the shortcut of IIS.

### Run a program from default home directory:-

1. Open the hard disk partition where the OS is loaded.
2. Open the directory(folder) Inetpub.
3. Open the directory wwwroot.
4. Create a new directory like "myasp", under "wwwroot"
5. Write some ASP code and save the file as "demo.asp" in "myasp" directory.
6. Open your web browser and type "http://localhost/myasp/demo.asp", to view demo.asp

### Run a program from any other directory:-

Virtual directory.

- It is also possible to execute asp files from any other location other than of home directory by creating a virtual directory.
- A virtual directory is an alias to any physical directory on the hard drive which contains the asp pages.
- User can access that physical directory using its virtual directory name. At the time of the web page request, IIS maps the virtual directory name

specified by the client in the URL of the requested page to its actual physical directory name.

- We can use some files in more than one web sites running on the same server by storing them inside a same virtual directory.

### Create a Virtual Directory:-

Suppose the default home directory for the asp applications is C:\Inetpub\Wwwroot\. But we want to store the asp application at D:\mywebsite\. The directory structure is shown below.

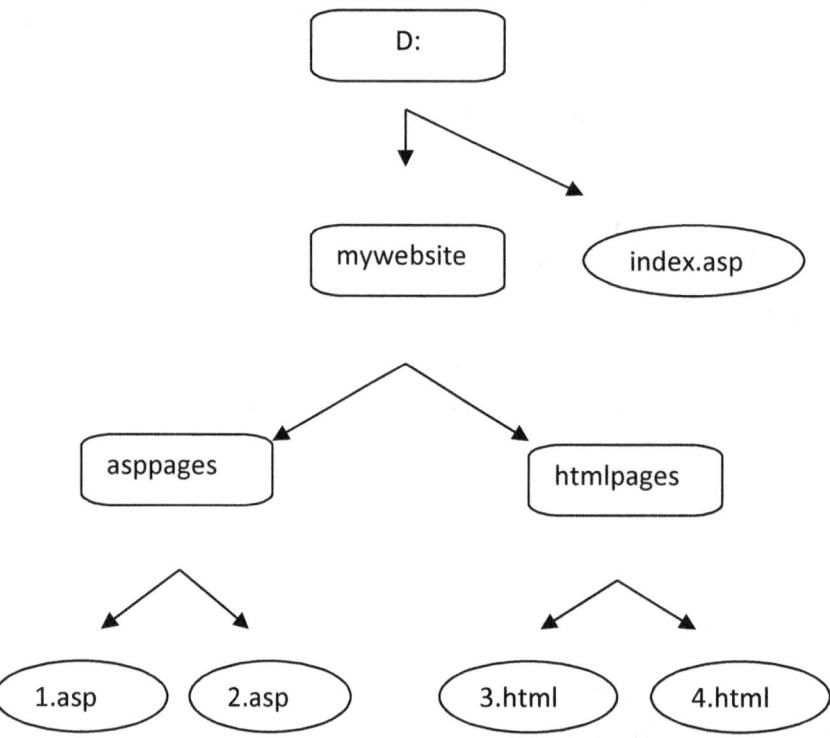

Here we have distributed all the web pages of our web site into two sub directories, "asppages" and "htmlpages" inside the "mywebsite" directory. So now we should create a virtual directory corresponding to D:/mywebsite/. If we want user to access our web site by the name "demoweb" then it is the virtual directory name.

1. Open control panel.

2. Open administrative tools.

3. Open Internet information services.

4. Expand websites category

5. Right click on "default web site"->New->Virtual Directory->Next.

6. Alias is the name of the virtual directory. Enter "demoweb"->Next.

7. Select the physical directory by using Browse. Here it is "D:/mywebsite/"

8. Next->Finish.

9. Expand default web site category. You will see "demoweb" listed in it.

10. Click on "demoweb". On right side of the portion, you will see two folders "asppages" and "htmlpages" and index.asp

11. Right click on index.asp -> Browse.

12. You will see the output of index.asp in a default web browser.

13. Suppose you want to see the output of 1.asp, and then double click on "asppages". You will see the list of files available inside "asppages" in the right side of the IIS window. 14. Select any file->Right click on it->Browse

It is also possible to run the files directly from the browser after creating virtual directory by specifying the URL in the address bar of the browser.

1:- run index.asp

http://localhost/demoweb/index.asp

IIS Server will map "demoweb" to "D:/mywebsite/" as per the information of "demoweb" virtual directory. It will find index.asp inside "D:/mywebsite/" and will run it.

2:-run 2.asp

http://localhost/asppages/2.asp

3:- run 4.html

http://localhost/htmlpages/4.html

## 5.5 ASP SYNTAX

An ASP file contains HTML tags and server side scripting(VBScript) statements. We already discussed the syntax of HTML tags in chapter 3. <% is used to indicate the stating of the server side script and %> is used to indicate the ending of the server side script. The default scripting language is VBScript. We can also change the scripting language by using,

<%@ language="javascript"%>

We have already discussed some examples of server side scripts in chapter 4.

5.1.asp (ASP syntax)

| | |
|---|---|
| ```<HTML><BODY><%Response.Write ("Dynamic WebDevelopment")Response.Write "<BR>"Response.Write "Dynamic Web Development"Response.Write "<BR><B><I>ASPProgramming"%></BODY></HTML>``` | OUTPUT:-<br><br>*Dynamic Web Development*<br>*Dynamic Web Development*<br>***ASP Programming*** |

Response object is a built-in object of ASP. Write is its method. Response.Write is used to write something to output. It can be a text or HTML tags which represent image, tables etc. in above program, we can see various ways to use Response.Write with HTML tags. Write method accepts a string as an argument and it interprets the string as a HTML statement. So wherever we place a HTML tag in argument of Write method, we will get that specific effect on the output.

Response.Write "<BR>" breaks a line.

Response.Write "<BR><B><I>ASP Programming" breaks a line and will set the bold and italics effect for rest of the text.

**Note: -** never break an ASP statement in two lines.

con1.open "Provider=Microsoft.Jet.OLEDB.4.0;Data Source="

& Server.MapPath(".") & "\data.mdb"

This is invalid and it will generate a runtime error.

If we want to display a statement in more than one lines, we use the special line break character "_" (the underscore)

con1.open "Provider=Microsoft.Jet.OLEDB.4.0;Data Source=" _

& Server.MapPath(".") & "\data.mdb"

If our text editor automatically wraps a statement in two or more lines then it is valid.

## 5.6 ASP VARIABLES

ASP variables are used to store information. There are four categories of ASP variables according to the scope of them.

108

**Local Variable:-** A variable declared inside a procedure. It is not possible to access this variable outside the procedure in which it is declared.

**Global Variable:-** A variable not declared inside any procedure is globally available in the .asp file in which it is declared.

It is also possible to access a variable across multiple asp files by declaring them as session variables or application variables.

**Session Variable:-** A session variable is used to store information across multiple asp pages accessed by a single user. It is normally used to keep track of user's activities on a web site, to remember username and login time of a user,to count total number of currently logged in users etc. ASP's Session object is used to manage session variables.

**Application Variable:-** An application variable is used to store information across all the users who access the web site. These variables are global to entire web application and common for all the users. It is normally used to keep information about total number of visitors, information about the connection to the database etc.

## 5.7 DIFFERENCE BETWEEN VBSCRIPT AND JAVASCRIPT

| No | VBScript | JAVAScript |
|----|----------|------------|
|    |          |            |
| 1 | VBScript is developed by Microsoft | JAVAScirpt is developed by Netscape Communications. |
| 2 | VBScript is the default server side scripting language in ASP based applications. | JAVAScript is not default scripting language in ASP based applications. |
| 3 | VBScript is mainly used as a server side scripting language. | JAVAScript is mainly used as a client side scripting language. |
| 4 | VBScript is case insensitive. | JAVAScript is case sensitive. |
| 5 | VBScript is syntactically similar to Visual Basic. | JAVAScript is syntactically similar to JAVA. |
| 6 | VBScript supports the concept of properties. | JAVAScript does not support the concept of properties. |
| 7 | VBScript is less object oriented. | JAVAScript is more object oriented. |
| 8 | VBScript does not support optional procedure argument | JAVAScript supports optional procedure arguments. |
| 9 | VBScript has powerful native MsgBox and InputBox functions for dialog/user input support. | JAVAScript does not have native functions for dialog/user input support. It depends on the object model.(window.alert, confirm, prompt or WScript.echo, popup) |
| 10 | Only Internet Explorer supports VBScript as a client side scripting language. | Mostly all popular browsers support JAVAScript as a client side scripting language. |

## 5.8 ASP OBJECTS

ASP supports two types of objects.

- **Built In Objects (Intrinsic objects):-**
  ASP supports various intrinsic built-in objects which are always available to use. We don't need to create them manually. Using ASP objects, we can collect information from the client, send information to the client, collect information about the web server, manage session and application variables etc.

Here is the summary of ASP objects and various functionalities supported by them.

| Object Name | Purpose |
| --- | --- |
| Request | Obtain information from user (browser). |
| Response | Provide information to user. |
| Session | Associate a specific web page request with a specific user. |
| Application | Store and retrieve global information across all the users. |
| Server | Obtain information about the server |
| ObjectContext | Integrate ASP with other objects |
| ASPErr | Manage errors & error reporting |

Each object has some information associated with it.

**Properties:-**Attributes or characteristics of an object. Properties describe current state of an object. We need to set the properties according to our requirements because the behavior of an object depends on its properties. Some properties are the read-only properties. Some are write-only and some are both.

Example:- in VB, TextBox has various properties like Text,MultiLine,PasswordChar etc. In ASP, for a connection object, ConnectionString, ConnectionTimeOut are the properties.

**Methods:-**a method Describes things that we can do with an object. It is a Procedure that acts on an object.

Example:- in VB, Textbox has a SetFocus method. In ASP, for a connection object, Open, Close are the methods.

**Events:-**Similar to a method but it is triggered by a human action or external system action.

Example:- in VB TextBox has a Change event, in ASP,for an application object, Application_OnStart() is an event.

**Collections:-**Set of related information to an object in key–value pairs with index numbers starting from 1.this is not similar to a 2-D array. The main difference is that we can change the size of the collection any time. There is a possibility to search the value by giving its corresponding key.

Example:- in VB, ListBox has a List property which is an indexed set of list items. There are some methods are available through which we can insert or remove an item from any index. In ASP, Form,QueryString etc. are collections of Request Object.

We will study all the ASP objects in detail with their properties, methods, events and collection in chapter 6.

- **ActiveX Objects:-**
    An instance of the object must be explicitly instantiated before it can be used.

These are the additional components we can use while working with ASP technology.

Some components are required to be installed on the server before we can use them. Mostly these components are COM or DCOM based, can be developed using different technologies.

Examples:-
Adrotator,BrowserCapabilities,FileSystemObject,Connection,RecordSet etc.

## 5.9 REVIEW QUESTIONS

1. What is the ASP technology? Why we need to use such technology in develpmenet of web sites?

2. What are the advantages and disadvantages of ASP.

3. Explain the ASP model.

4. What is virtual directory?

5. Explain types of variables in ASP.

6. Explain ASP objects.

7. Differentiate VBScript and JAVAScript.

8. Differentiate HTML request and ASP request.

# Chapter 6:- Communication With The User

## 6.1 ASP BUILT-IN OBJECTS

We discussed ASP Objects in chapter 5. Page no (XXX). In this chapter we will study each of the following objects in detail. Each object has some properties, methods, events and collections.

| Object Name | Purpose |
|-------------|---------|
| Response | Provide information to user. |
| Request | Obtain information from user (browser). |
| Application | Store and retrieve global information across all the users. |
| Session | Associate a specific web page request with a specific user. |
| Server | Obtain information about the server. |

## 6.2 RESPONSE OBJECT

When a user (browser) sends a request of data from the web server, the server responds with either required data or an error. The response of the server contains two parts.

(1)   The response header contains information about the data.
(2)   The response body contains the actual data.

For HTML page requests, the server returns the content of HTML pages.

For ASP page requests, the server uses the Response Object to generate response data in HTML form.

Example:-

Suppose user has requested the result of roll no "101". The server has an asp file which collects result data of "101" from database. Response object will be used to represent the output of the execution of asp file into HTML.

### 6.2.1 Methods of Response Object:-

• Write:- (Preparing output)
     This method writes text and HTML tags to the response. All the content which we write using Write method will be visible on the client browser in the form of server's response. Similarly BinaryWrite method is used to write non-textual data to the browser.

Syntax  :- Response.Write text/HTML data

Example:- Response.Write "<B>Hello World!</B>"

- End:- (Stopping execution)
     This method stops further execution of the requested file(script or HTML content). The server will not execute any code that follows the Response.End statement.

Syntax  :- Response.End

Example:- Response.End

- Clear:- (Erasing the current output)
     This method clears(removes) the current response content. So if server needs to erase the response after generating it, Response.Clear method is useful.

Syntax  :- Response.Clear

Example:- Response.Clear

- Flush:- (Sending the current output to the client)
server stores the response data into its local buffer if the buffer property of Response object is set to true. At the end of the execution, server sends the entire response to the client. So if the page is long (in terms of execution and output data), client will need to wait for long time. So it is better to send response to the client in small pieces. Response.Flush sends any buffered response to the browser.after it server can continue the execution with rest of the page.

Syntax  :- Response.Flush

Example:- Response.Flush

- AddHeader:-(Redirection from the client)
This method adds additional header information to the response. One useful information that we can add is the URL at which client browser will move after specifying seconds. This process is known as redirection. We must need to call this method before any call of Flush method. This is the automatic redirection from the client.

Syntax  :- Response.AddHeader "Type of Information","Information"

Example:- Response.AddHeader "Refresh","5;URL=abc.asp"

             After 5 seconds, Client browser will move to abc.asp

- Redirect:-(Redirection at the server)
AddHeader method is used to do redirection to a URL after specific amount of time the Client browser Redirect method is used to redirect immediately. The

server does not execute any code which is written after Redirect call. The call to Redirect method gives an error if Buffer property of Response object is not set to true. The redirect starts the execution of a new page. So it clears all the previously generated response. This is the automatic redirection at the server.

Syntax :- Response.Redirect "URL"

Example:- Response.AddHeader "abc.asp"

6.1.asp (Respose.Write)

| <HTML><br><BODY><br><%<br>Response.Write "Hello World!"<br>Response.Clear<br>Response.Write "<B>Today is " &<br>formatDateTime(Date,vbLongDate)<br>Response.Write "<HR>"<br>%><br></BODY><br></HTML> | **OUTPUT:-**<br><br>Today is Monday, May 30, 2011 |
| --- | --- |

## 6.2.2 Properties of Response Object:-

• Buffer:-
If this property is set to true then server buffers the response content until the page execution completes or until the execution of Response.Flush or Response.End. Then server sends the response to the client. Default value is false. It is required to write the Response.Buffer statement at the beginning of the ASP file, otherwise it gives an error.

Example :- Response.Buffer = True

(True is the keyword and so it does not required to place it inside the quotations).

• CacheControl:-
This property is used to informa the proxy server whether it needs to cache the current page or not. If sets to "Public", proxy server caches the page and provides directly to the client without sending request to the server. Default is "Private" (No caching)

Example :- Response.BufferCachControl = "Public"

- CharSet:-

This property sets the character set the browser should use to display the page. Default character set is ISO-LATIN-1.

Example :- Response.CharSet="ISO-8859-1"

- ContentType:-

This property sets the type of content the page contains. This information is useful to search engine. The value is the representation of type and subtype.

Example  :-        Response.ContentType="text/HTML"
            Response.ContentType="image/GIF"
            Response.ContentType="image/JPEG"
            Response.ContentType="text/plain"

- Expires:-

This property informs the proxy server (or browser), the amount of duration (in minutes) for which it is allowed to cache the page(for which Expires property is set). After that duration, it will need to send a request to the server for the fresh response of the page. if a user requests for the same page before the page expires, the cached version of that page is displayed.

Example  :- Response.Expires=-1          will never cache

        Response.Expires=10     will expire after 10 minutes

- ExpiresAbsolute:-

This property informs the proxy server (or browser), the date and time when the cached version of a page will expire. if a user requests for the same page before the page expires, the cached version of that page is displayed. And if after the specified date and time then client will send a new request to the server for the fresh copy of that page.

Example  :- Response.ExpiresAbsolute=#July 11,2011 16:00:00#

6.2.3 Collections of Response Object:-

## 6.2.3 Collections of Response Object:-

Response object has only one collection which is Cookies. We will discuss Cookies in 6.5

## 6.2.4 Events of Response Object:-

There is no event of Response Object.

## 6.3 REQUEST OBJECT

The request object contains the information that the client(browser) sends to the server.

The server and the client treat every request as a new request so with every request, client sends some information (client's ip address, content of all fields of form, browser type information etc.) to the server. The request object is used to get these information for the use during the execution of requested page.

Example:-

Suppose user sends the roll no to the server. Request object is used to get the roll no which user has sent to the server from the page which displays the result based on the roll no.

### 6.3.1 Methods of Request Object:-

*   BinaryRead:-
    This method reads the non-textual data sent to the server. It requires the number of bytes we want to read as an argument.

### 6.3.2 Properties of Request Object:-

*   TotalBytes:-
    This property returns the total number of bytes the browser has sent in the request.

Example  :-  Response.Write Request.TotalBytes

> It writes total number of bytes, browser sent in the request.

### 6.3.3 Collections of Request Object:-

*   Form:-
    This collection is used when browser submits the form using post method.

This collection contains values of all the fields of a form entered by the user.

Example  :-  Response.Write Request.Form("studentname")

This prints  the content of the textbox studentname.

*   QueryString:-

This collection is used when browser submits the form using get method.

This collection contains values of all the fields of a form appened into the requested URL because of the use of get method.

Example :- Response.Write Request.QueryString("studentname")

This prints the content of the textbox studentname.

We will discuss Form and QueryString collections in detail in 6.5

- ClientCertificates:-
  This collection contains Information ( of authentication) sent by the browser required for the request to a secured site which is based on HTTPS (secure version of HTTP)

- ServerVariables:-
  This is the set of environment variables sent by the browser with every new request. Here is the list of some of the important server variables.

| Name of the ServerVariable | Example | Purpose |
|---|---|---|
| LOCAL_ADDR | 127.0.0.1 | IP Address Of Server |
| REMOTE_ADDR | 127.0.0.1 | IP Address Of Client |
| SERVER_NAME | Localhost | Host Name |
| SERVER_PORT | 80 | Useful In TCP/IP |
| SERVER_SOFTWARE | Microsoft IIS/5.1 | Iis Version |
| SERVER_PROTOCOL | HTTP/1.1 | Http Version |
| REQUEST_METHOD | GET | Get Or Post |
| QUERY_STRING | ?name=hkm&dept=ce | Querystring If Get |
| HTTPS | Off | Secure HTTP |
| PATH_TRANSLATED | F:\dwd\asp\3.ASP | Entire Physical Path |
| URL | /dwd/3.ASP | Only File Name |
| SCRIPT_NAME | /dwd/3.ASP | Only File Name |
| APPL_PHYSICAL_PATH | F:\dwd\asp\ | Physical Path |
| HTTP_USER_AGENT | Opera/9.27 | Client Browser |

We can access individual server variable using

Request.ServerVariables("HTTP_USER_AGENT")

6.2.asp (Print All Server Variables)

| | OUTPUT:- |
|---|---|
| ```<br><HTML><br><BODY><br><TABLE BORDER=1><br><%<br>Dim v<br>For Each v in Request.ServerVariables<br>Response.Write "<TR>"<br>Response.Write "<TD>" & v<br>Response.Write "<TD>" &<br>Request.ServerVariables(v)<br>Next<br>%><br></BODY><br></HTML><br>``` | It prints all the server variables in a tabular form. |

- Cookies:-

We will discuss Cookies collection of Request object in 6.5

## 6.4 STATE MANAGEMENT

It is not possible to develop an interactive program without using any variables. In chapter 4, we used variables. But they are local variables accessible in a page or within a function in which they were declared. For interactive web sites, we need to maintain some information between pages. So we need some way through which we can keep some variables global to our web site. Keeping information accessible across multiple web pages is the state management.

**Example1:-**

Assume we want to implement following functionality. A user enters username and password in login page. If username and password are correct then user will be allowed to access other pages. We also want to display the username on the beginning of all the pages. To implement the above functionality, we need to keep username accessible through out all the pages visited by a user. If two users are accessing our web site at a time then server should remember both the username independently.

**Example2:-**

We want to keep track of how many users have visited our web site. So we create a counter and whenever a new user requests for a web page, server should increment the counter by one. So at any instance counter contains the total number of users visited our web site so far. To implement the above functionality, server should keep one variable accessible to all the users. Only one copy of counter variable should be maintained by the server.

**Example3:-**

After successfully login of a user, we want to display the total number of users who are online right now. For this also, server needs to maintain a global variable. with each successful login, server should increment that variable by 1 and with each successful logout, server should decrement that variable by 1. There are two places to maintain the state, either on the server or on the client. Here is a list of state management techniques used in ASP programming.

1. Using Form and QueryString Collections:-

Form and QueryString collections of Request object are available if we post the form using Post and Get methods respectively. The content of fields of a form is accessible in another file to which you submit the form. These variables reside in server's RAM.

2. Using Cookies:-

Using cookies, we can store and retrieve values stored on the client machine's hard disk. So cookies are accessible in future even after closing the web site.

3. Session Variables:-

Session variables are created using Session object. These variables are created with each new user request and remain accessible through out the visiting period of that user. These variables reside in server's RAM.

4. Application Variables:-

Application variables are created using Application object. These variables are shared variables accessible by all the users. Only one copy of these variables is maintained by the IIS. These variables reside in server's hard disk

## 6.5 ASP COOKIES

### 6.5.1 What is a cookie?

- A cookie is a small piece of information provided by the server and stored as a small file on client's hard disk.
- A cookie has the form of key and value representation.
  Firstname=hardik

  In this example, firstname is the cookie name and hardik is the value.

  We can also create cookies in HTML using <META> tag.

- Cookies are simply some extra information in the header field of the response or request.
- Response object has a Cookies collection which is used to set(create, modify or delete) cookies.Request object has a Cookies collection which is used to access the values of cookies.

120

- After we create cookies, browser sends the cookies to the server each time we make a request to the site which created those cookies. Each time the browser requests a URL, browser checks for the cookies associated with that URL. If browser finds any such cookies then it sends all those cookies with the request header.
- Cookies are secure because browser sends only those cookies to the site (server – if the server is hosting only one site) which are actually created by that site. So no other site can read cookies created by some other sites.
- We can set the expiration date and time after which browser expires the cookies and make them no longer valid. If we don't set the date and time, browser stores cookies in the memory of the client and it expires them when we close the browser.
- If cookies contains any sensitive information(username, password) which needs to be securely delivered to the server then it is possible to send the encrypted form of cookies using HTTPS.

## 6.5.2 Creating a cookie

- The response object's Cookies collection is used to create a cookie or modify or delete any existing cookie.
  Response.Cookies("firstname") = "Hardik"

  Response.Cookies("lastname") = "Molia"

     These two statements create two cookies firstname and lastname on client machine.

- The same statement can be used to modify the existing cookies. To delete a cookie, we assign "" (blank) value to the cookie.


  Response.Cookies("firstname") = ""

     This statement deletes firstname cookie from the client machine.

- We can set the expiration date using following statement


  Response.Cookies("firstname").Expires = #July 15,2011#

## 6.5.3 Retrieving a cookie

- With each new web request, browser automatically inserts all the cookies generated by the site of which the new request belongs in the header section. We can access the cookies using the Cookies collection of Request object. Request.Cookies("firstname") returns the value of the cookie firstname

  Response.Write  Request.Cookies("firstname")
     This way we can print the value of the cookie firstname.

## 6.3.asp (Create Cookies)

| | OUTPUT:- |
|---|---|
| ```<br><HTML><br><BODY><br><%<br>Response.Write "<H3>Page 1</H3>"<br>Response.Cookies("username") = "Hardik Molia"<br>Response.cookies("logintime") = Time<br>Response.write "<BR><A HREF=6.4.asp>Next Page</A>"<br>%><br></BODY><br></HTML><br>``` | **Page 1**<br><br>Next Page |

## 6.4.asp (Retrieving Cookies)

```
<HTML>
<BODY>
<%
Response.Write "<H3>Page 2</H3>"
Response.cookies("logintime") = Time
Response.Write "Hello " & Request.Cookies("username")
Response.Write "<BR>You Are Logged in Since:- " &
Request.Cookies("logintime")
%>
</BODY>
</HTML>
```

**OUTPUT:-**

**Page 2**

Hello Hardik Molia
You Are Logged in Since:- 6:58:56 AM

### 6.5.4 Multilevel Cookie

- It is also possible to create a sub cookie under a main cookie. If we want to include "firstname", "lastname" and "city" information under a main cookie "user" then we define multiple values inside a single key as shown below.

Response.Cookies("user")("firstname") = "Hardik"

Response.Cookies("user")("lastname") = "Molia"

Response.Cookies("user")("city") = "Rajkot"

6.5.asp (Creating Multi level Cookies)

```
<HTML>
<BODY>
<%
Response.AddHeader "Refresh","10;URL=6.6.asp"
Response.Cookies("user")("firstname")="Hardik"
Response.Cookies("user")("lastname")="Molia"
Response.Cookies("user")("country")="India"
Response.Cookies("user")("city")="Rajkot"
Response.Write "Cookies created...<BR>Wait For 10 Seconds..."
%>
</BODY>
</HTML>
```

**OUTPUT:-**

Cookies created...
Wait For 10 Seconds...

6.6.asp (Retrieving Multi level Cookies)

```
<HTML>
<BODY>
<%
Response.Write "Name:-" & Request.Cookies("user")("firstname") & "<BR>"
Response.Write "Surame:-" & Request.Cookies("user")("lastname") & "<BR>"
Response.Write "Country:-" & Request.Cookies("user")("country") & "<BR>"
Response.Write "City:- " & Request.Cookies("user")("city")  & "<BR>"
%>
</BODY>
</HTML>
```

**OUTPUT:-**

Name:-Hardik
Surame:-Molia
Country:-India
City:- Rajkot

### 6.5.5 Reading all the cookies

It is difficult to write the code to read all the cookies by specifying their keys. As Cookies is a collection, we can use for each... statement to read all the cookies in iterative manner as shown below.

6.7.asp (Reading All Cookies)

```
<HTML>
<BODY>
<%
Dim x,y
For Each x in Request.Cookies
  If Request.Cookies(x).HasKeys Then
    For Each y in Request.Cookies(x)

          Response.Write x & ":" & y & "=" & Request.Cookies(x)(y) &
"<BR>"
    Next
Else
    Response.Write x & "=" & Request.Cookies(x) & "<BR>"
End If
Next
%>
</BODY>
</HTML>
```

**OUTPUT:-**

username=Hardik Molia
logintime=6:58:56 AM
user:firstname=Hardik
user:lastname=Molia
user:country=India
user:city=Rajkot

124

Request.Cookies(x).HasKeys statement is used to find out whether the cookie Request.Cookies(x) has sub keys or not. If Request.Cookies(x).HasKeys returns true then the inner loop prints all the sub cookies.

## 6.6 ASP FORMS

- While designing a web site, we often need to get some information from the client and according to that information, we decide further execution.

- For a registration process of an email account, there will be a page which has a form containing various fields like name, surname, country, city, password etc. based on the information provided by the user, another page creates an email account.

- For an online result delivery, a form is used to get the branch, semester and roll no information. Based on this information, server generates the result.

- In this section we will study how to get content of a form (information entered by the user) into another asp page.

- To submit a form, we use ACTION and METHOD attributes of the <FORM>. When we want to submit a form to the server, we need to specify the URL of the file which server will execute on receiving the form. ACTION attribute specifies the URL of the file which contains the code to get the form content and further processing. METHOD attribute specifies how client browser sends the form information to the server. Tow possible methods are POST and GET. Default method is GET.

## GET:-

- In GET method, browser generates a QueryString which is one of the components of URL (Discussed in chapter 1.) QueryString contains the name and values of all the fields of submitted form.

- So if we post a form using GET method, browser generates a querystring and then it appends it to the URL specified in the ACTION attribute. So finally the new URL contains the filename and the querystring.

- If we are sending some sensitive data (passwords) then this method is not preferable because all the contents will be visible as a part of the URL.

- Request object's querystring collection is used to retrieve the form content.

## 6.8.asp (A Form)

```
<HTML>
<BODY>
<FORM ACTION=6.9.ASP METHOD=GET>

NAME:- <INPUT TYPE="TEXT" NAME="T1"></INPUT><BR><BR>
PASSWORD:- <INPUT TYPE="PASSWORD" NAME="P1"></INPUT><BR><BR>
DIPLOMA:- <INPUT TYPE="CHECKBOX" NAME="C1"></INPUT><BR><BR>
DEGREE:- <INPUT TYPE="CHECKBOX" NAME="C2"></INPUT><BR><BR>

MALE <INPUT TYPE="RADIO" NAME="R1" VALUE="MALE"></INPUT>
FEMALE<INPUT TYPE="RADIO" NAME="R1"
VALUE="FEMALE"></INPUT><BR><BR>

ADDRESS:- <TEXTAREA NAME="T2" ROWS=5
COLS=25></TEXTAREA><BR><BR>

BRANCH:- <SELECT NAME="B1">
<OPTION>CIVIL
<OPTION>MECH.
<OPTION>I.T.
<OPTION>E.C.</SELECT>
<BR><BR>

CELL CONNECTION:-
<SELECT NAME="CN1" SIZE=4 MULTIPLE>
<OPTION>!DEA
<OPTION>VODAPHONE
<OPTION>AIRTEL
<OPTION>BSNL</SELECT><BR><BR>

<INPUT TYPE = "SUBMIT">
<INPUT TYPE = "RESET">
<INPUT TYPE = "BUTTON" VALUE="Exit">

</FORM>
</BODY>
</HTML>
```

**OUTPUT:-**

NAME:- HARDIK MOLIA

PASSWORD:- *********

DIPLOMA:- ☑

DEGREE:- ☑

MALE ◉ FEMALE ○

RAJKOT

ADDRESS:-

BRANCH:- I.T. ▾

CELL CONNECTION:- IDEA
VODAPHONE
AIRTEL
BSNL

[ Submit ] [ Reset ] [ Exit ]

Assume that the physical path of 6.8.asp is  E:\dwd\ch 6\book example\6.8.asp

And the virtual directory corresponding to E:\ is myweb

So when user submits the above form, browser generates a querystring as shown below.

http://localhost/chj6/dwd/ch%206/book%20example/6.8.asp?T1=HARDIK+MOLIA&P1=QSXRFVYHN&C1=on&C2=on&R1=MALE&T2=RAJKOT&B1=I.T.&CN1=%21DEA

http://localhost/chj6/dwd/ch%206/book%20example/6.8.asp is the URL of the file specified in the ACTION attribute of the form. According to base-64 encoding, " " in URL are replaced with %20.

'?' indicates the beginning of the querystring.

'&' separates each component of the querystring.

'+' indicates a ' ' (space) in querystring.

T1=HARDIK+MOLIA indicates the value of T1 field is "HARDIK MOLIA"

P1=QSXRFVYHN indicates the value of P1 field is "QSXRFVYHN" and so on...

Because of the GET method, browser sends the above URL to the server. Now server finds various values using Request object's QueryString collection.

6.9.asp (Access form data using QueryString Collection)

```
<HTML>
<BODY>
<%
Response.Write "<BR>Name:- " & Request.QueryString("T1")
Response.Write "<BR>Password:- " & Request.QueryString("P1")
Response.Write "<BR>Diploma:- " & Request.QueryString("C1")
Response.Write "<BR>Degree:- " & Request.QueryString("C2")
Response.Write "<BR>Gender:- " & Request.QueryString("R1")
Response.Write "<BR>Address:- " & Request.QueryString("T2")
Response.Write "<BR>Branch:- " & Request.QueryString("B1")
Response.Write "<BR>Cell:- " & Request.QueryString("CN1")
%>
</BODY>
</HTML>
```

**OUTPUT:-**

Name:- HARDIK MOLIA
Password:- QSXRFVYHN
Diploma:- on
Degree:- on
Gender:- MALE
Address:- RAJKOT
Branch:- I.T.
Cell:- IDEA

## POST:-

- POST method is used to submit the form invisibly without generating QueryString. So this method is more suitable for the submission of sensitive information.

- Request object's Form collection is used to retrieve the form content.

Replace METHOD = POST in example 6.8. Now browser sends the form values in invisible way to the server. Server uses Form collection of Request object to find various values.

6.10.asp (Access form data using Form Collection)

```
<HTML>
<BODY>
<%
Response.Write "<BR>Name:- " & Request.Form("T1")
Response.Write "<BR>Password:- " & Request.Form("P1")
Response.Write "<BR>Diploma:- " & Request.Form("C1")
Response.Write "<BR>Degree:- " & Request.Form("C2")
Response.Write "<BR>Gender:- " & Request.Form("R1")
Response.Write "<BR>Address:- " & Request.Form("T2")
Response.Write "<BR>Branch:- " & Request.Form("B1")
Response.Write "<BR>Cell:- " & Request.Form("CN1")
%>
</BODY>
</HTML>
```

**OUTPUT:-**

Name:- HARDIK MOLIA
Password:- QSXRFVYHN
Diploma:- on
Degree:- on
Gender:- MALE
Address:- RAJKOT
Branch:- I.T.
Cell:- !DEA

## 6.7 FORM VALIDATION

We discussed the importance of client side and server side scripts in chapter 4.2 we discussed an example of registration process in which we needed to verify the data entered by the user at both client side as well as at server side. (Refer 4.2).

In this topic, we will discuss how to write client side validation code and server side validation code with ASP programming.

### 6.7.1 Client Side Validation:-

It is not possible to access the elements of the page at client side using Response object. We must have to use Document object. The document object is used to access all the elements of a HTML page. we can use document object to set the background color of the page using document.bgcolor = "#FFAA3D"

Suppose the name of the form is "form1". Name of the textbox is "t1". We can use document.form1.t1.value to access the value of the textbox "t1"

Client side validation code is required to be executed by the browser before the submission of the form to the server. If browser gets success in all the validation then browser submits the form using document.form1.submit where "form1" is the name of the form.

Because of the client side scripts, we can use msgbox and inputbox to interact with the user.

We can use javascript as a scripting language also. Here is the example in which vbscript is used to write client side validation routine.

While writing code for client side validation, instead of inserting a submit button, we should use a simple button. On click event of the button, we call the validation routine. The routine validates all the fields step by step, gives error message if any invalid data is present otherwise submits the form.

6.11.asp (Client Side Validation)

```
<HTML>
<HEAD>
<SCRIPT LANGUAGE="VBSCRIPT">
Sub validate()
      If Document.form1.t1.value = "" Then
            MsgBox "Pls Enter Your Name..."
            Exit Sub
      End If

      If isnumeric(Document.form1.t1.value) = true Then
            MsgBox "Name must not be a number..."
```

```
                    Exit Sub
            End If


            If len(Document.form1.p1.value) < 5 Then
                    MsgBox "password length is too small..."
                    Exit Sub
            End If


            If Document.form1.c1.checked = false and Document.form1.c2.checked =
false Then
                    MsgBox "You must have completed atleast diploma..."
                    Exit Sub
            End If


            If Document.form1.cl1.selectedindex = -1  Then
                    MsgBox "Pls select your cell phone connection..."
                    Exit Sub
            End If


            Document.form1.Submit
End Sub
</SCRIPT>


<BODY>
<FORM ACTION="6.10.asp" METHOD="POST" NAME="form1">
<TABLE>
<TR><TD>NAME:- <TD><INPUT TYPE="TEXT" NAME="T1">
<TR><TD>PASSWORD:- <TD><INPUT TYPE="PASSWORD" NAME="P1">

<TR><TD>DIPLOMA:- <TD><INPUT TYPE="CHECKBOX" NAME="C1"
value="diploma">
<TR><TD>DEGREE:- <TD><INPUT TYPE="CHECKBOX" NAME="C2"
value="degree">
<TR><TD></TD>
<TD>MALE <INPUT TYPE="RADIO" NAME="R1" VALUE="male"
CHECKED></INPUT>
FEMALE <INPUT TYPE="RADIO" NAME="R1" VALUE="female"></INPUT>

<TR><TD>ADDRESS:- <TD><TEXTAREA ROWS=5 COLS=25 NAME="a1">
</TEXTAREA>

<TR><TD>BRANCH:- <TD><SELECT NAME="B1" >
<OPTION>CIVIL<OPTION>MECH.<OPTION>I.T.<OPTION>E.C.</SELECT>

<TR><TD>CELL PHONE CONNECTION:-<TD> <SELECT NAME="Cl1" SIZE=4
```

```
MULTIPLE>
<OPTION>!DEA<OPTION>VODAPHONE<OPTION>AIRTEL<OPTION>BSNL</SEL
ECT>

<TR><TD><INPUT TYPE="button" VALUE="send" onclick="validate()">
</FORM>
</BODY>
</HTML>
```

**OUTPUT:-**

NAME:-        HARDIK MOLIA

PASSWORD:-    ••

DIPLOMA:-     ☐

DEGREE:-      ☐

          MALE ◉ FEMALE ○

ADDRESS:-

BRANCH:-      CIVIL ▾

                IDEA
                VODAPHONE
CELL PHONE CONNECTION:-   AIRTEL
                BSNL

send

VBScript ✕

password length is too small...

OK

In above example, validation is a subroutine which validates the form fields. For each validation rule, we have written a if... else.. Code. If the condition is will be true, a message box will appear containing the error message and Exit Sub will stops further validation.

Here is the list of validation rules, which we have implemented in client side validation.

1- Name must not be empty.

2 – Name must not be a number.

3 – Minimum password length is 5.

4 – At least select checkbox of diploma(A user must have selected at least Diploma checkbox.

5- Compulsory selection of cell phone connection.

We can use the built-in functions to implement various validation rules as shown in the above example.

If there is no validation error, all the if conditions will become false and so execution control will never enter into any if block and so at the end of the subroutine, document.form1.submit will submit the form to 6.10.asp

Server will accept the data and will display the result as shown in example 6.10

### 6.7.2 Server Side Validation:-

In server side validation, we submit the form using either the Get method or the Post method. When server receives the form content, it validates the data by executing some code. And so this type of validation is server side validation.

Consider the form of example 6.8. The only change is in the ACTION attribute. Replace the <FORM> tag with following.

<FORM ACTION=6.12.ASP METHOD=GET>

6.12.asp (Server Side Validation)

```
<HTML>
<BODY>
<%
Dim name,password,diploma,degRee, add, branch,gEnder,cell

name = Request.Form("t1")
password = Request.Form("p1")
diploma = Request.Form("c1")
degree = Request.Form("c2")
add = Request.Form("t2")
branch = Request.Form("b1")
gender = Request.Form("r1")

If name = "" Then
     Response.Write "Name is blank..."
     Response.Write "<INPUT TYPE=BUTTON VALUE=BACK
onClick=history.back()>"
     Response.End
End If

If isnumeric(name) = true Then
```

```
        Response.Write "Name is numeric..."
        Response.Write "<INPUT TYPE=BUTTON VALUE=BACK
onClick=history.back()>"
        Response.End
End If


If password <> "12345678" Then
        Response.Write "invalid password."
        Response.Write "<INPUT TYPE=BUTTON VALUE=BACK
onClick=history.back()>"
        Response.End
End If


Response.Write "<BR>Name    :- " & name
Response.Write "<BR>Password:- " & password
Response.Write "<BR>Diploma :- " & diploma
Response.Write "<BR>Degree  :- " & degree
Response.Write "<BR>Address :- " & add
Response.Write "<BR>Branch  :- " & branch
Response.Write "<BR>Gender  :- " & gEnder
Response.Write "<BR>CellConnections :- "

For Each item in Request.Form("cn1")
        Response.Write item & "<BR>"
Next
%>
</BODY>
</HTML>
```

**OUTPUT:-(Name field is blank)**

Name is blank... BACK

**OUTPUT:- (all fields are validated successfully.)**

Name :- HARDIK MOLIA
Password:- 12345678
Diploma :- YES
DegRee :-
AddRess :-
Branch :- I.T.
Gender :- MALE
CellConnections :-
IDEA
AIRTEL

134

In server side validation, we stores values of form fields into variables. We check for the validity using various built-in functions as shown above. Here is the list of validation rules, which we have implemented in server side validation.

1- Name must not be empty.

2 – Name must not be a number.

3 – Password must be "12345678"

For Each item in Request.Form("cn1")

        Response.Write item & "<BR>"

Next

We enabled multiple selections in list box by setting MULTIPLE attribute. The above loop prints all the cell connections selected by the user.

history is a built-in navigation object. we used history object's back() to movie to one page back on click event of a button.

onClick=history.back()

## 6.8 APPLICATION OBJECT

- Application object is used to store information global to all the users who are accessing the web site.

- Application object manages global data so to avoid data inconsistency; Application object supports locking facility of data by using Lock and Unlock methods. So while one user is accessing the Application Object's data, no other user will allow to access.

Examples:-

- Suppose we want to keep track of how many users have visited our web site. So we create a counter using Application object to make it globally accessible to all the users. Whenever a new user requests for a web page of our site, server increments the counter by one. So at any instance counter contains the total number of users visited our web site so far.

- A dynamic web site mostly has database connectivity. It is not required to open the connection to the database every time a new user requests for a page. So it is possible to establish the connection to the database using a global Application object and make it accessible to all the users.

### 6.8.1 Collections of Application Objects:-

- Contents:-

Application object's Contents collection is the set of global values in key-value form. These values are common for all the users and stored on server's hard disk.

### Creating/Modifying Application Variable:-

Application("Counter") = 1

Creates an application variable Counter and initialize it to 1.

Same statement can be use to modify any existing application variable.

### Accessing Application Variable:-

We can access the Contents collection in several ways. Here is the possible ways to write the value of the variable "Counter" from the Contents collection of Application object.


Response.Write Application.Value("Counter")

Response.Write Application("Counter")

Response.Write Application.Contents("Counter")

Response.Write Application.Contents(1)

### Removing Application Variable:-

Application("Counter") = ""

Removes the counter variable from the Contents collection of Application object.

We can also remove a variable using Remove method.

Application.Contents.Remove("Counter")          removes "Counter" variable

Application.Contents.Remove(1)          removes the variable which at index 1

Application.Contents.RemoveAll()          removes all the variables from Contents

- StaticObjects:-


To use objects at application level, this collection is useful. StaticObjects is the collection of all the objects defined using <OBJECT> tags at application level.

## 6.8.2 Methods of Application Objects:-

- Lock:-

Application object is globally available to all the users. While one user is accessing the application object it is required to lock the application object so no other user can access it. This will avoid data inconsistency due to simultaneous access of application object.

Lock method locks the application object(makes inaccessible to other users) until we executes an Unlock method, until page execution ends or until the page times out. We should lock the application object to add or update application variables.

- Unlock:-

It unlocks the application object (makes accessible to other users) if it is locked. We should always unlock the application object immediately after completing the add or update operation on application object.

- Contents.Remove(index) and Contents.RemoveAll are also methods of Application objects discussed in 6.8.1

## 6.8.3 Events of Application Objects:-

- Application_OnStart()

This event occurs before the server serves the first user request of our web site.this event does not occur with any subsequent user requests. The event occurs when the web application gets its first request only.

This event is useful to write some initialization code. We can create the hit counter or can establish the connection to database which will be useful through out the life of web application.

- Application_OnEnd()

This event occurs when our web application stops. We can use this event to destroy any global data stored in application object.

We can write code for both the events in a special configuration file global.asa (Discussed in 6.10)

## 6.9 SESSION OBJECT

- When we work with a desktop application, we execute it, use it and close it. This is a session. The desktop computer knows who we are. It knows when we started the application, what we worked with it and when we closed it.

But in web application, there is one problem; the web server gets lots of requests simultaneously. The web server does not whether the user is requesting 1st time or he is the user who has been served just before few minutes. This is because of the stateless HTTP protocol.

- ASP solves the problem by creating a unique session object for each user. A session object is used by the server to identify all the users uniquely using the unique SessionID assigned to all the users.

- When a user requests for the first time, ASP engine creates a Session object for the user and assigns an unique SessionID. When a user closes the browser window or session times out, web server destroys the corresponding session object.

- The session object is used to identify each user uniquely. It is used to store user information like user's personal information, preferences, login time, logout time, activities etc.

- The difference between application object and session object is that, application object is global to all the users while each user gets a unique session object which is not accessible by any other users.


## 6.9.1 Collections of Session Objects:-


- Contents:-
  Session object's Contents collection is the set of values in key-value form. These values are accessible in all the pages for the user for which they were created.

Creating/Modifying Session Variable:-

Session("Name") = "Hardik Molia"

Session("login") = Now

Creates two session variables("Name" and "login").

Same statement can be use to modify any existing session variable.

### Accessing Session Variable:-

We can access the Contents collection in several ways. Here is the possible ways to write the value of the variable "name" from the Contents collection of Session object.
Response.Write Session.Value(("Name")

Response.Write Session("Name")

Response.Write Session.Contents("Name")

Response.Write Session.Contents(1)

**Removing Session Variable:-**

Session("Name") = ""

Removes the Name variable from the Contents collection of Session object

We can also remove a variable using Remove method.

Session.Contents.Remove("Name")          removes "Name" variable

Session.Contents.Remove(1)          removes the variable which at index 1

Session.Contents.RemoveAll()          removes all the variables from Contents

- StaticObjects:-

To use objects at session level (accessible to all the pages visited by the user in whose session objects are created), this collection is useful. StaticObjects is the collection of all the objects defined using <OBJECT> tags at session level.

### 6.9.2 Methods of Session Objects:-

- Abandon:-
    This method destroys the session of the user. The user of whose session has been deleted is no longer allowed to access the web site.

- Session.Abandon
    Destroys the session of the user for whom the above statement is executed.

- Contents.Remove(index) and Contents.RemoveAll are also methods of Application objects discussed in 6.8.1

### 6.9.3 Events of Session Objects:-

- Session_OnStart()
    This event occurs when a user requests for any page of the web site first time. This event is used to intializae some values for the user's session which is accessible through out the user's visit. This event is a part of the global.asa. For each new user, the event is occurred.

- Session_OnEnd()
    This event occurs when user's session ends (abandoned or times out). This event is useful to destroy any data stored at session level.

We can write code for both the events in a special configuration file global.asa (Discussed in 6.10)

### 6.9.4 Properties of Session Objects:-

* SessionID:-
   IIS assigns a unique Session ID to all the sessions at any instance. This is the read only property.

Server identifies a particular user by his sessionID.

* LCID:-
   This property sets or returns the locale identifier. This property is used to set the locale information to provide dates and currency values in different formats based on the locale (location information).

session.LCID = 2057

response.write formatcurrency(500)               According to English – U. K.

* CodePage:-
   This property sets or returns the code page number which is used to specify the character set. Default is 1252(American English).

* TimeOut:-
   This property sets or returns the timeout period for the user Session in minutes. If the user does not do any activity with the current page belonging to his session, the session will end. Default is 20 minutes in IIS 5 and 10 minutes in IIS 5.

### 6.10 global.asa FILE

* The Global.asa file is an optional configuration file that contains objects, variables, and methods which are accessible in all the pages of an ASP application.

* We use scripting language to write code of global.asa file. The global.asa must be saved in the root directory(or virtual directory) of the ASP application.

* using global.asa, we can inform the application and session object what to do when application starts, when session starts, when session ends and when application ends.

- global.asa contains four events. We can write code for these events in scripting language.

## Events in Global.asa

- Application_OnStart() – When application starts

This event occurs when the 1st user requests for the first page in an ASP application. This event also occurs when web server is restarted or after the editing of global.asa. This event is useful to set global information accessible across all the users.

The session_OnEnd() event occurs immediately after this event.

- Session_OnStart() – When Session Starts

This event occurs every time a new user requests his first page from the ASP application. This event is useful to set information accessible across all the pages unique to a user.

- Session_OnEnd() –  When Session Ends

This event occurs when user's session ends (abandoned or times out). This event is useful to destroy any data stored at session level. This event is useful to destroy all the information related to a user's session.

- Application_OnEnd() – When Application Stops

This event orrcurs when we stop the web server, after the end of last user's session. This event is used to remove any application level variables and objects from the memory of the server.

(Application Hit Counter)Global.asa

```
<SCRIPT LANGUAGE="VBSCRIPT" RUNAT="SERVER">
Sub Application_OnStart()
Application("visitors")=0
End Sub

Sub Application_OnEnd()
Application("totvisitors")=0
End Sub
</SCRIPT>
```

6.13.asp

| | OUTPUT:- |
|---|---|
| ```<HTML><BODY><%Application.LockApplication("visitors") = Application("visitors") + 1Application.UnlockResponse.Write "Total Hits:- " &Application("visitors")%></BODY></HTML>``` | Total Hits:- 15 |

Description:-

- Application_OnStart() initialize an application variable "visitors" by 0. With each page request, server locks the application object, increments the application variable "visitors" by 1 and unlocks the application object.

- Application object is common for all the users so at any instance, Application object's Contents collection's "visitors" variable contains the total number of users visited so far.

(Count the total no. of online users)

Global.asa

```
<SCRIPT LANGUAGE="VBSCRIPT" RUNAT="SERVER">

Sub Session_OnStart()
Application.Lock
Application("totalusers")=Application("totalusers")+1
Application.UnLock
End Sub

Sub Session_OnEnd
Application.Lock
Application("totalusers")=Application("totalusers")-1
Application.UnLock
End Sub

</SCRIPT>
```

6.14.asp

| | OUTPUT:- |
|---|---|
| ```<br><HTML><br><BODY><br><%<br>Response.Write "Total Online Users:- " &<br>Application("totalusers")<br>Session.Abandon<br>%><br></BODY><br></HTML><br>``` | Total Online Users:- 22 |

## Description:-

- When a new user requests for a page first time, server executes Session_OnStart() and when his session expires, server executes Session_OnEnd()

- Session_OnStart() increments the application variable "totalusers" by 1.(To increment total number of online users by 1 because of the beginning of new session)

- Session_OnEnd() decrements the application variable "totalusers" by 1.(To decrement total number of online users by 1 because of the ending of the session)

- It is required to lock and unlock the Application object in Session events to avoid data inconsistency.

- Server executes the Session_OnEnd() only when server ends the session using either Abandon method or because of the Session time out.

## 6.11 SERVER OBJECT

The server object is used to perform various activities related with the server only.

### 6.11.1 Properties of Server Objects:-

- ScriptTimeOut:-

Suppose user requests an asp file. Server starts execution of that file, but the execution goes into execution of an infinite loop. User is waiting for the

143

reply. Server is executing also but it is not possible for the user to get the reply. One another possibility is the situation of deadlock over resources.

To avoid such possibility of infinite executions, this property sets the duration in seconds, after which a script will time out if processing is not complete. The default value is 90 seconds.

We should set this value to a large value if our site remains busy(large number of users who want to access at a time) or is having very large pages.

### 6.11.2 Methods of Server Objects:-

• CreateObject:-

This method is used to create an ActiveX object. The argument is the valid ProgID(specifies the classname).

Set adrot=Server.CreateObject("MSWC.AdRotator")

It is compulsory to use Set keyword while assigning objects. The above statement creates an object adrot of AdRotator component.

We will see the use of CreateObject in chapter 7.1

• Execute:-

This method is used to execute another ASP file from the currently executing file. After the execution of another file, ASP engine returns back to the caller asp file and resumes the execution from the next statement.

Server.Execute("file2.asp")
Executes file2.asp

(Execute Method)

6.15.asp

| ```
<HTML>
<BODY>
<%
Response.Write "<BR>Hello World.This is file
6.15.asp"
%>
</BODY>
</HTML>
``` | OUTPUT:- <br><br> Hello World.This is file 6.15.asp |
|---|---|

144
```

6.15execute.asp

| | OUTPUT:- |
|---|---|
| ```<HTML><BODY><%Response.Write "<BR>This is file 6.15execute.asp"Response.Write "<BR>Let's execute 6.15.asp"Server.Execute("6.15.asp")Response.Write "<BR>Back to the file6.15execute.asp"%></BODY></HTML>``` | This is file 6.15execute.aspLet's execute 6.15.aspHello World.This is file 6.15.aspBack to the file 6.15execute.asp |

- Transfer:-

     This method is used for the server side redirection. It transfers the excution control to another asp file. The control does not return back to the caller asp file. This method is more efficient than Response object's redirect method.

Server.Transfer("file2.asp")
Transfers the control to the file2.asp

(Transfer Method)

6.16.asp

| | OUTPUT:- |
|---|---|
| ```<HTML><BODY><%Response.Write "<BR>Hello World.This is file6.16.asp"%></BODY></HTML>``` | Hello World.This is file 6.16.asp |

6.16transfer.asp

| | OUTPUT:- |
|---|---|
| ```<HTML><BODY><%Response.Write "<BR>This is file6.16transfer.asp"Response.Write "<BR>Let's transfer control to``` | This is file 6.16transfer.aspLet's transfer control to 6.16.aspHello World.This is file 6.16.asp |

145

```
6.16.asp"
Server.Transfer("6.16.asp")
Response.Write "<BR>Back to the file
6.16transfer.asp"
%>
</BODY>
</HTML>
```

- MapPath:-

This method is used to perform virtual path to physical path conversion. We often need to specify the URLs (of images, of database files, of other web pages etc). It is required to specify the full physical path instead of giving only filename. After writing absolute physical paths, If we change the root directory to any other location then all the physical paths will become useless. To avoid such situation we specify only the file name and MapPath method finds the physical path from the virtual path with reference to the virtual directory.

(Transfer Method)

6.17.asp

Assume that 6.17.asp is stored at E:\dwd\ch 6\examples\6.17.asp

| <HTML><br><BODY><br><%<br>Response.Write Server.MapPath(".")<br>Response.Write "<BR>"<br>Response.Write Server.MapPath("..")<br>Response.Write "<BR>"<br>Response.Write Server.MapPath("/")<br>Response.Write "<BR>"<br>Response.Write Server.MapPath("\")<br>Response.Write "<BR>"<br>Response.Write Server.MapPath("new\a.asp")<br>%><br></BODY><br></HTML> | **OUTPUT:-**<br><br>E:\dwd\ch 6\examples<br>E:\dwd\ch 6<br>c:\inetpub\wwwroot<br>c:\inetpub\wwwroot<br>E:\dwd\ch 6\examples\new\a.asp |
| --- | --- |

"."             Represents current directory

".."            Represents one level up directory from the current directory.

"/" and "\"   Represent default directory for IIS application

- HTMLEncode:-

This method is used to replace ampersands (&) angle brackets (< and >) and other non text characters with special token so browser can display these characters as the original characters. This method is useful if we want to display the string as a sequence of characters instead of treating it as a HTML statement.

- URLEncode:-

This method is used to apply URL Encoding rules to a string as per the base-64 encoding scheme. This method is used to replace all the special symbols (space, : , /, # etc) with the percent sign(%) concatenated with the ASCII value of the replaced character in hexadecimal.

(HTMLEncode and URLEncode)

6.18.asp

```
<HTML>
<BODY>
<%
Response.Write "<B>Bold Text</B>"
Response.Write "<BR><BR>"
Response.Write Server.HTMLEncode("<B>Bold Text</B>")
Response.Write "<BR><BR>"
Response.Write Server.URLEncode("www.xyz.com\my docs\p1.asp")
%>
</BODY>
</HTML>
```

**OUTPUT:-**

**Bold Text**

<B>Bold Text</B>

www%2Exyz%2Ecom%5Cmy+docs%5Cp1%2Easp

## 6.12 EXAMPLES

**Example 6.19** (Result Calculation)

The program has two files.

6.19form.asp contains the code for the form having fields for name,rollno, marks of physics, chemistry and biology, submit and reset button. All fields and arranged in a tablular form for proper layout.

6.19result.asp accepts the form data. It calculates the percentage and grade and displays in a tabular form.

6.19form.asp

```
<HTML>
<BODY>
<FORM ACTION=6.19result.asp METHOD=POST>
<TABLE>

<TR><TD>ENTER NAME:-
<TD><INPUT TYPE="TEXT" NAME="SNAME">
</INPUT><BR>

<TR><TD>ENTER ROLL NO:-
<TD><INPUT TYPE="TEXT" NAME="SROLL" SIZE=2>
</INPUT><BR><BR>

<TR><TD>ENTER MARKS OF PHYSICS:-
<TD><INPUT TYPE="TEXT" NAME="MPHY" SIZE=2>
</INPUT><BR><BR>

<TR><TD>ENTER MARKS OF CHEMISTRY:-
<TD><INPUT TYPE="TEXT" NAME="MCHE" SIZE=2>
</INPUT><BR><BR>

<TR><TD>ENTER MARKS OF BIOLOGY:-
<TD><INPUT TYPE="TEXT" NAME="MBIO" SIZE=2>
</INPUT><BR><BR>

<TR><TD><INPUT TYPE = "SUBMIT">
<TD><INPUT TYPE = "RESET">
</TABLE>
</FORM>
</BODY>
</HTML>
```

**OUTPUT:-**

ENTER NAME:-           HARDIK MOLIA

ENTER ROLL NO:-        12

ENTER MARKS OF PHYSICS:-    85

ENTER MARKS OF CHEMISTRY:-  80

ENTER MARKS OF BIOLOGY:-    77

[ Submit ]               [ Reset ]

## 6.19result.asp

```
<HTML>
<BODY>
<%
Dim sname,sroll,physics,chemistry,bio,total,per,grade
sname = Request.Form("SNAME")
sroll = Request.Form("SROLL")
physics = CInt(Request.Form("MPHY"))
chemistry = CInt(Request.Form("MCHE"))
biology = CInt(Request.Form("MBIO"))

total = physics+chemistry+biology

per = total/3

If per>=70 Then
grade ="Distinction"
ElseIf per>=60 Then
grade ="First Class"
ElseIf per>=50 Then
grade ="Second Class"
ElseIf per>=36 Then
grade ="Pass Class"
Else
grade ="Fail"
```

149

```
End If

With Response
.Write "<TABLE BORDER=1>"
.Write "<TR><TD COLSPAN=2 ALIGN=CENTER><H1>RESULT</H1>"
.Write "<TR><TD>NAME <TD>" & sname
.Write "<TR><TD>ROLL NO <TD>" & sroll
.Write "<TR><TD>PHYSICS <TD>" & physics
.Write "<TR><TD>CHEMISTRY <TD>" & chemistry
.Write "<TR><TD>BIOLOGY <TD>" & biology
.Write "<TR><TD>PERCENTAGE <TD>" & Round(per,2) & "%"
.Write "<TR><TD>GRADE <TD>" & grade
End With
%>
</FORM>
</BODY>
</HTML>
```

**OUTPUT:-**

# RESULT

| NAME | HARDIK MOLIA |
|------|--------------|
| ROLL NO | 12 |
| PHYSICS | 85 |
| CHEMISTRY | 80 |
| BIOLOGY | 77 |
| PERCENTAGE | 80.67% |
| GRADE | Distinction |

Note:- QueryString or Form collection is a set of strings. So we used CInt() function to convert marks from string to number. It is always necessary to use the exactly same name of form field while accessing its value in another page using either QueryString or Form collection.

**Example 6.20** (User login using session)

6.20.asp

This example implements simple login functionality. Valid username is "Hardik" and valid password is "1234567". On entering valid username and password, it displays another page which creates two session variables. ("username" – to store user name and "logintime"- to store the time at which the user logged in). The page contains one logout link. After logging out, another page displays a message and the amount of duration for which the user was logged in (in minutes) using DateDiff function as shown below.

6.20loginform.asp

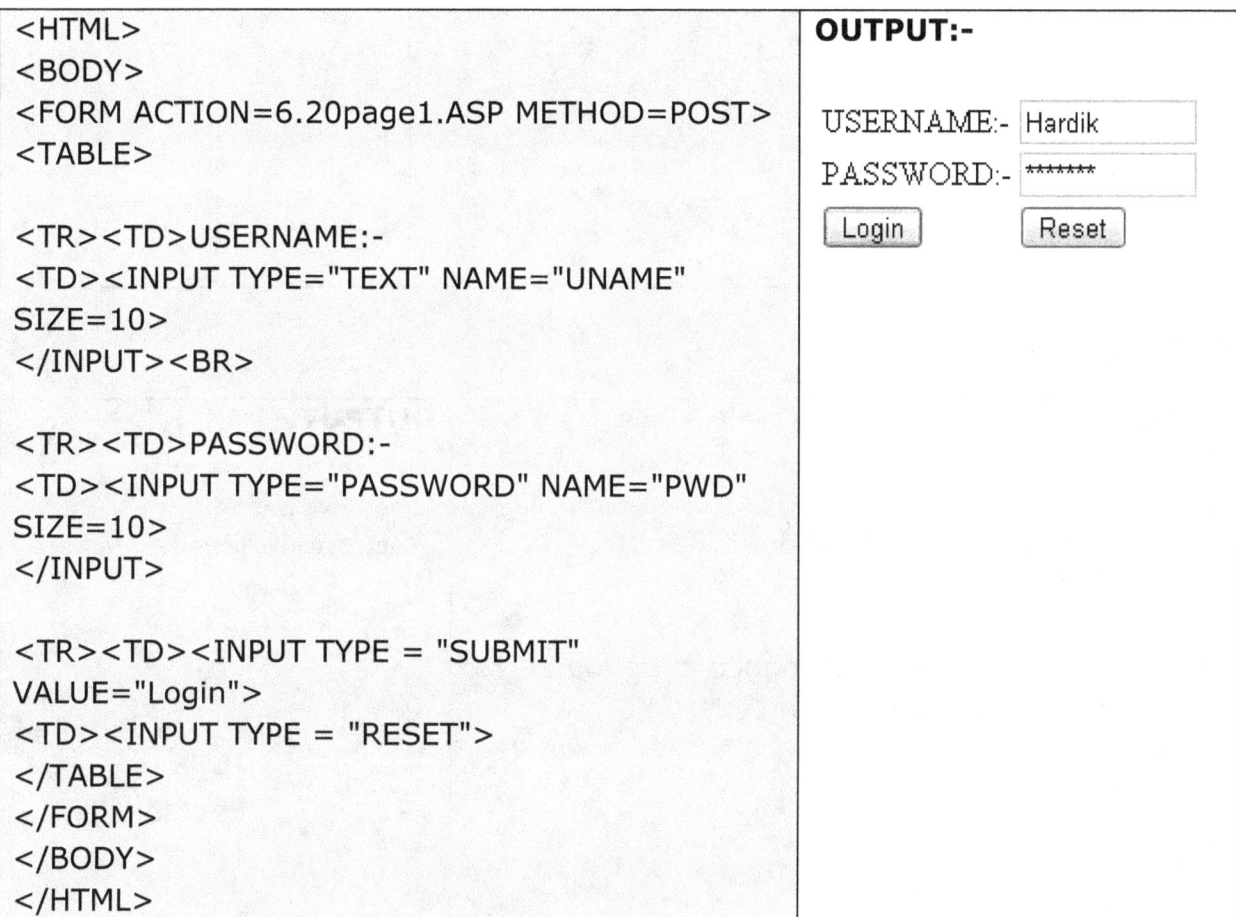

```
<HTML>
<BODY>
<FORM ACTION=6.20page1.ASP METHOD=POST>
<TABLE>

<TR><TD>USERNAME:-
<TD><INPUT TYPE="TEXT" NAME="UNAME"
SIZE=10>
</INPUT><BR>

<TR><TD>PASSWORD:-
<TD><INPUT TYPE="PASSWORD" NAME="PWD"
SIZE=10>
</INPUT>

<TR><TD><INPUT TYPE = "SUBMIT"
VALUE="Login">
<TD><INPUT TYPE = "RESET">
</TABLE>
</FORM>
</BODY>
</HTML>
```

**OUTPUT:-**

USERNAME:- Hardik

PASSWORD:- *******

[ Login ]    [ Reset ]

6.20page1.asp

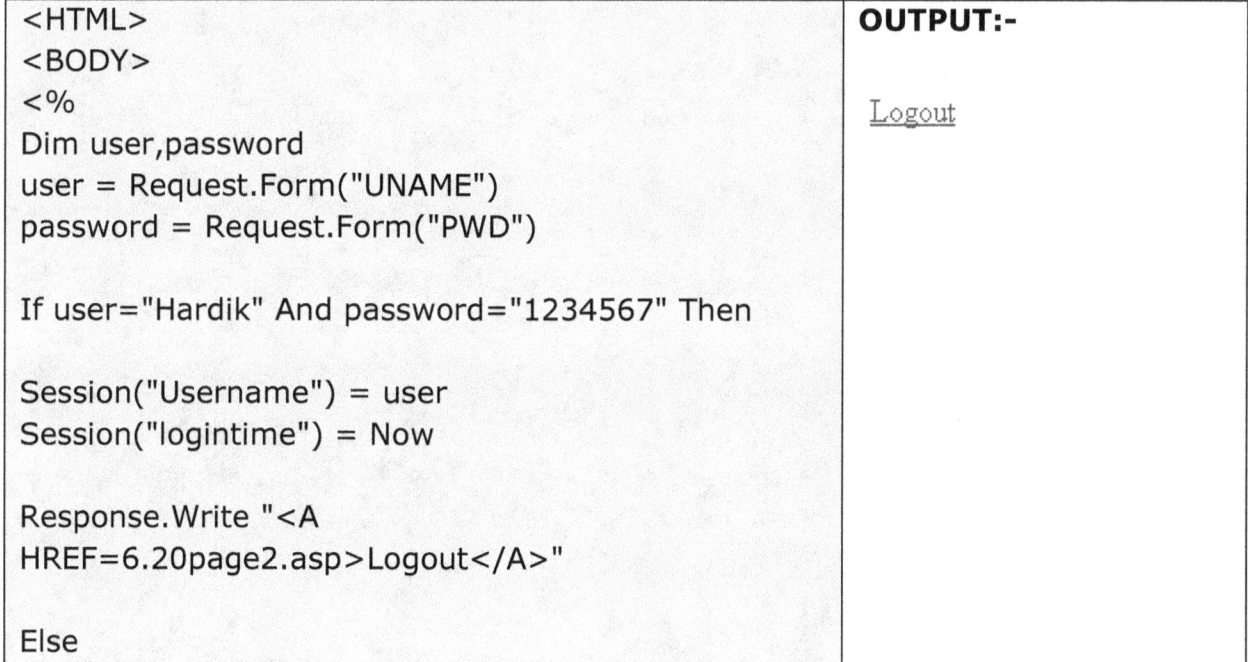

```
<HTML>
<BODY>
<%
Dim user,password
user = Request.Form("UNAME")
password = Request.Form("PWD")

If user="Hardik" And password="1234567" Then

Session("Username") = user
Session("logintime") = Now

Response.Write "<A
HREF=6.20page2.asp>Logout</A>"

Else
```

**OUTPUT:-**

Logout

151

```
Response.Write "Invalid username or password."
Response.Write "<INPUT TYPE=BUTTON
VALUE=BACK onClick=history.back()>"

End If

%>
</FORM>
</BODY>
</HTML>
```

6.20page2.asp

```
<HTML>
<BODY>
<%
Response.Write "<H3>Good Bye " &
Session("username")

Response.Write "<BR>Your logged in period:- " &
DateDiff("n",Session("logintime"),Now) & "
Minutes."

%>
</FORM>
</BODY>
</HTML>
```

**OUTPUT:-**

Good Bye Hardik
Your logged in period:- 6 Minutes.

## 6.13 REVIEW QUESTIONS

1. Write short notes on following ASP objects.
     Response, Request, Application, Session and Server

2. Explain following methods of Response object with simple examples.
     Write, End, Clear, Flush, AddHeader, Redirect

3. Explain following properties of Response object with simple examples.
     Buffer, CacheControl, CharSet, ContentType, Expires, ExpiresAbsolute

4. Explain following collections of Request object with simple examples.
     Form, QueryString, ClientCertificates, ServerVariables, Cookies

5. Explain state management in ASP.

6. Explain Cookies.

7. Differentiate GET method and POST method.

8. Differentiate Request's Cookies and Response's Cookies.

9. Explain following collections of Application object.
     Contents, StaticObjects

10. Explain how Session is useful in state management.

11. Explain global.asa

12. Explain following methods of Server object.
     CreateObject, Execute, Transfer, MapPath, HTMLEncode, URLEncode

13. Differentiate Response's Redirect and Server's Transfer.

14. Differentiate Server's Execute and Server's Transfer.

15. Explain client side and server side form validations.

# Chapter 7:- Database Handling With Asp

## 7.1 COMPONENTS OF ASP

An asp component is a collection of code used to provide some additional functionality in the web designing. Microsoft installs several components with ASP in IIS. We will study two of them. (AdRotator and BrowserCapabilities).

### 7.1.1 AdRotator Component:-

i hope you have seen many web sites with advertisement banners. The advertisement changes every 10 or 20 seconds. And with every new request, you will get new advertisement in random order. If you click on any of the advertisement, you will be redirect to corresponding web site. The AdRotator component is used to implement this functionality through ASP.

To use AdRotator in our web page, we must first create an AdRotator schedule text file. It contains two parts.

### General Section:-

This section contains the name of an ASP file to which browser redirects when a user clicks on the advertisement. It also specifies the width, height and border information of the advertisement.

### Detail Section:-

This section contains the list of images for advertisements, corresponding URLs, alternate text (for image disabled browsers) and numbers that specify the frequencies of advertisements.

### Example 7.1 (AdRotator Demo.)

adverts.text (AdRotator Schedule File)

```
REDIRECT adrotator.asp
BORDER 2
width 400
height 100
*
rajkot.gif
www.rajkot.com
Rajkot
50
it.gif
www.itjobs.com
IT Jobs
30
wallpapers.gif
```

```
www.wallpapers.com
wallpapers
20
```

It is required to create a separate AdRotator schedule file, for each AdRotator we use in asp files.

The file starts with the general section.

| Field | Meaning |
|---|---|
| REDIRECT filename | If this filename is specified, when user clicks on any advertisements, the browser requests the file specified in this field. |
| WIDTH pixels | Width of the AdRoator. |
| HEIGHT pixels | Height of the AdRoator. |
| BORDER size | Border size of the AdRoator. |

* Indicates the end of the general section.

For each advertisement we provide three values.

1. URL of the advertisement image.

2. URL of the page/site which browser requests if user clicks on this particular advertisement.

3. Frequency of the advertisement.

AdRotator itself selects an advertisement from the advertisements specified in the schedule file randomly. But it maintains the frequency as per specified with each advertisement.

rajkot.gif

www.rajkot.com

Rajkot

50

The advertisement's image URL is rajkot.gif

If user will click on this advertisement, browser will request for www.rajkot.com

The frequency is 50. It means in 100 hits of the page containing AdRotator, you will get this advertisement 50 times. (Order is random)

Similarly there are two other advertisements.

7.1.asp (AdRotator)

| | OUTPUT:- |
|---|---|
| ```<HTML><BODY><%Set                    Adr                    =Server.CreateObject("MSWC.AdRotator")Response.WriteAdr.GetAdvertisement("adverts.txt")%></BODY></HTML>``` | 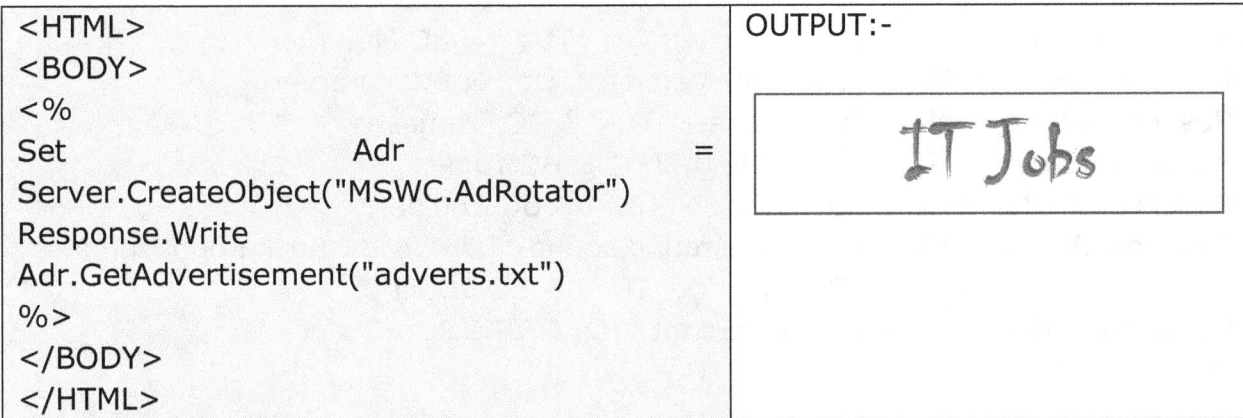 |

AdRotator is a part of MSWC (Microsoft Web Class) components. First of all we create an object Adr using server object's createobject method.

To attach an AdRotator schedule file with Adr, we use Adr's GetAdvertisement method with schedule file's URL in argument.

We put the Adr using Response.Write as shown in above example.

When we will run 7.1.asp many times, we will see different advertisements in AdRotator as per the adverts.text

## 7.1.2 BrowserCapabilities Component:-

We discussed how to get the name of the client browser using Request object's ServerVariables collection's HTTP_USER_AGENT key.

The BrowserCapabilities component is used to collect more information about he client browser. So using this information, server can decide what to produce as the response.

Using this component, the web server can find out whether the client browser supports tables or not. If not then server will generate the response without tables. Similarly for frames, cookies, vbscript, javascript etc.

7.2.asp

```
<HTML>
<BODY>
<%
Set BC = Server.CreateObject("MSWC.BrowserType")
Response.Write "The Client Browser:-" & Request.ServerVariables("HTTP_USER_AGENT")

Response.Write "<BR><BR><TABLE BORDER=1><TR><TH>FIELD<TH>VALUE"
Response.Write "<TR><TD>Name<TD>" & BC.Browser
```

```
Response.Write "<TR><TD>OS<TD>" & BC.Platform
Response.Write "<TR><TD>Version<TD>" & BC.Version
Response.Write "<TR><TD>Major Version<TD>" & BC.Majorver
Response.Write "<TR><TD>Minor Version<TD>" & BC.Minorver
Response.Write "<TR><TD>Frames<TD>" & BC.Frames
Response.Write "<TR><TD>Tables<TD>" & BC.Tables
Response.Write "<TR><TD>Cookies<TD>" & BC.Cookies
Response.Write "<TR><TD>Background sound<TD>" & BC.Backgroundsounds
Response.Write "<TR><TD>VBScript<TD>" & BC.VBScript
Response.Write "<TR><TD>JavaScript<TD>" & BC.JavaScript
%>
</BODY>
</HTML>
```

**OUTPUT:-**

The Client Browser:- Mozilla/4.0 (compatible; MSIE 6.0; Windows NT 5.1; FunWebProducts; .NET CLR 1.0.3705; .NET CLR 2.0.50727; InfoPath.2)

| FIELD | VALUE |
|---|---|
| Name | IE |
| OS | WinNT |
| Version | 6.0 |
| Major Version | 6 |
| Minor Version | 0 |
| Frames | True |
| Tables | True |
| Cookies | True |
| Background sound | True |
| VBScript | True |
| JavaScript | True |

We use server object's CreateObject method to create an object of BrowserType component of MSWC. We must have to use set keyword while creating an object.We can access various properties of the browser using object's (here it is BC)'s properties as shown in above example.

158

We can find out client browser's name, client's platform(os), browser's versions, whether the browser supports frames, tables, cookes, background sound, vbscript and java script etc.

If the client browser supports vbscript then BC.VBScript has True value. Similar for other information.

## 7.2 ACTIVEX DATA OBJECTS (ADO)

In dynamic web sites, we store data in database. We often need to access the data from our web pages according to the request of a user. We store results of students in a database. User enters a roll number and asp file finds his result from the database. There are several real life web applications where database access is required. Username and password verifications, bank transactions, railway reservation, movie reservation, online shopping etc. In this section we will study how to select, insert, update and delete information to and from any database using ADO.

Microsoft's ActiveX Data Objects (ADO) is a set of objects for accessing various databases. We can use ADO objects to access database without knowing how the database is implemented. The Microsoft Jet Database Engine is a software package on which Microsoft's database products were built

ADO works with both ODBC data sources and OLEDB data sources. ODBC is used to access relational databases while OLEDB is used to access relational as well as non relational databases.

ADODB is a database library that simplifies database related tasks. Before going through the programming, it is required to understand the ADO model.

In all our examples, we will use following table created in Access 2003. data.mdb

### students : Table

| Field Name | Data Type | Description |
|---|---|---|
| srollno | Number | ROLL NO |
| sname | Text | NAME AND SURNAME |
| saddress | Text | ADDRESS |
| scity | Text | CITY |
| spercentage | Number | PERCENTAGE |

### 7.2.1 ADO Objects:-

- **Connection:-** A connection is used to establish the connection with the database.

- **Command:-** A command object represents a SQL statement or stored procedure of a database.

- **Parameter:-** A parameter object represents the parameter of a stored procedure.

- **RecordSet:-** A RecordSet object is a group of records returned from the result of a SQL query or from the execution of a stored procedure.

- **Record:-** A record represents one record in the database, and contains a Fields collection. A RecordSet consists of a collection of Record objects.

- **Stream:-** A Stream object is used with RecordSet object to read and write a stream of bytes. Used to save data in XML format.

- **Field:-** A Field corresponds to a column in the database. Each Record contains many fields.

## 7.3 CONNECTION OBJECT

Before we can retrieve any data from a database, we have to create a connection to the database. The Connection Object is used to create and open connection to a data source.

Set objConnection=Server.CreateObject("ADODB.connection")

### 7.3.1 Properties of Connection Object:-

- **ConnectionString** property sets or returns the information of the connection.
conn.ConnectionString="Provider=Microsoft.Jet.OLEDB.4.0"

Provider property sets or returns the provider name for a Connection object.

| Provider Name | Provider |
|---|---|
| Microsoft.Jet.OLEDB.4.0 | Microsoft Jet databases (e.g. Access) |

| MSDAORA | Oracle databases |
|---------|------------------|
| MSDASQL | Microsoft OLE DB provider for ODBC |
| SQLOLEDB | Microsoft SQL Server |

- **ConnectionTimeout** property sets or returns the number of seconds, ASP engine waits while opening a connection. After this time ASP engine cancels the connection attempt and generates an error. Default is 15 seconds.

- **Version** property returns the ADO version number.

### 7.3.2 Methods of Connection Object:-

- **Open** method opens a connection to a data source. When the connection is open, we can execute commands to retrieve data.

- **Close** method closes the Connection established with the database.

- **Cancel** method cancels different tasks for each object.

- **BeginTrans CommitTrans RollbackTrans** methods are used with the Connection object to save or cancel changes made to the data source.

- **Execute** method executes a SQL statement, stored procedure, or provider-specific text. It returns a read only RecordSet object.

Example:-

```
<%
Set con1 = Server.CreateObject("ADODB.Connection")
con1.open        "Provider=Microsoft.Jet.OLEDB.4.0;Data        Source="        &
Server.MapPath(".") & "\data.mdb"
Set rs1 = Con1.Execute("select * from students)
'Statements to process RecordSet rs1
con1.Close
%>
```

Instead of giving an absolute path for data.mdb, we can use Server.MapPath method to get the path of current directory.

## 7.4 RECORDSET OBJECT

The ADO RecordSet object is used to store the data returned from the database as a result of an execution of SQL statement or a stored procedure.

Set objRecordSet=Server.CreateObject("ADODB.RecordSet")

When we first open a RecordSet, the current record pointer points to the first record and the BOF and EOF properties are False. If there are no records, the BOF and EOF properties are True.

### 7.4.1 Properties of RecordSet Object:-

- **BOF** property returns True (-1) if the position of the record pointer is before the first record, otherwise it returns False (0).

- **EOF** property returns True (-1) if the position of the record pointer is after the last record, otherwise it returns False (0).

- **BookMark** property sets or returns a bookmark to identify the current record of the RecordSet uniquely.

  If rs is the RecordSet object,
  Mybk = rs.Bookmark       bookmarks current record
  rs.Bookmark = Mybk       locates rs to the record pointed by Mybk

- **CursorType** property specifies how we can navigate records in RecordSet. It also specifies whether is it possible to see the database changes made by other users or not.

In ADO there are 4 different cursor types defined:

| Constant | Value | Description |
|---|---|---|
| adOpenForwardOnly | 0 | Default. We can move the record pointer in forward direction only. |
| adOpenKeyset | 1 | All types of record pointer movement are possible.<br><br>Only changes and deletions by other users are visible. Additions by other users are not visible. |
| adOpenDynamic | 2 | All types of record pointer movement are possible.<br><br>Additions, changes, and deletions by other users are visible |

| adOpenStatic | 3 | All types of record pointer movement are possible. |
|---|---|---|
| | | Additions, changes, and deletions by other users are not visible. |

- **CursorLocation** Property sets the location of the cursor(client side or server side)

adUseServer

when maintaing a cursor server side, the responsibility of handling the data generated by a select query is on server. On server side, we can use all the available cursor types according to our requirements.

adUseClient

when maintaining a cursor client side, the responsibility of handling the data generated by a select query is on client and so this reduces the load on server. Only the adOpenStatic cursor is available with the client side cursor.

- **LockType** property specifies the type of locking scheme while editing a record of a RecordSet. This property is required to set properly to avoid data inconsistency in database.

| Constant | Value | Description |
|---|---|---|
| adLockReadOnly | 1 | Read-only records. Editing is not possible. |
| adLockPessimistic | 2 | locks a record immediately after editing |
| adLockOptimistic | 3 | locks records only when calling update method |
| adLockBatchOptimistic | 4 | Locks all records until the execution of BatchUpdate completes. |

- **RecordCount** property returns the number of records in a RecordSet object. This property will return -1 for a forward-only cursor and the actual count for other cursors. The RecordSet object must be open when reading the value of this property.

### 7.4.2 Methods of RecordSet Object:-

Open method opens a database element that gives you access to records in a table, the results of a query, or to a saved RecordSet.

objRecordSet.Open source,conn,cursortype,locktype

| Parameter | Description |
|---|---|
| source | Specifies a data source. It can be SQL statement, stored procedure, a table name or a command object. |
| conn | A connection string or a Connection object to the database. |
| cursortype | Specifies the type of cursor to use when opening a RecordSet object. Default is adOpenForwardOnly |
| locktype | Specifies the type of locking on a RecordSet object. Default is adLockReadOnly |

Instead of using Connection object's Execute method we can also use RecordSet object's Open method to execute a SQL statement. The Connection's Execute method always returns a read only RecordSet. So if we want to update the database, we must have to use RecordSet's Open method.

```
<%
Set con1 = Server.CreateObject("ADODB.Connection")
con1.open      "Provider=Microsoft.Jet.OLEDB.4.0;Data      Source="      &
Server.MapPath(".") & "\data.mdb"
Set rs1 = Server.CreateObject("ADODB.RecordSet")
rs1.Open "Select * from students",con1,3,3
con1.Close
%>
```

rs1.Open "Select * from students",con1,3,3
where con1 is the connection object, 3 specifies adOpenStatic cursor and 3 specifies adLockOptimistic lock type.

Note:- remember that rs1 is the RecordSet object created in above example.

- **MoveFirst** method sets the record pointer of a RecordSet to the first record.
  rs1.MoveFirst

- **MoveNext** method sets the record pointer of a RecordSet to the next record from the current record. If the current record is the last record, this will set the EOF to True.
  rs1.MoveNext

- **MovePrevious** method sets the record pointer of a RecordSet to the previous record from the current record. If the current record is the first record, this will set the BOF to True.
  rs1.MovePrevious

- **MoveLast** method sets the record pointer of a RecordSet to the last record.
  rs1.MoveLast

- **Move** method moves the record pointer in a RecordSet object to a specific location.

164

objRecordSet.Move numrec,start

| Parameter | Description |
|-----------|-------------|
| numrec | Specifies number of records to move from current record in forward or reverse direction. If this parameter is set to 3, the record pointer moves 3 records forward. If this parameter is set to -3, the record pointer moves 3 records backward |
| start | Where to start. The value can be<br><br>0=starts from the current record<br><br>1=starts from the first record (in forward direction)<br><br>2=starts from the last record (in reverse direction) |

rs1.Move 4,0      moves record pointer of rs1 four records in forward direction from the current record.

rs1.Move 4,2      invalid. It is not possible to movie record pointer in forward direction from the last record.

- **AddNew** method creates a new record for an updateable RecordSet object. After we call this method, the new record will be the current record. There are several ways we can add a new record.

Suppose we want to insert a record in students table. If rs1 is the recordset, we can use any of the following methods to insert a record.

| 1st method:- | 2nd method:- |
|-----------|-------------|
| rs1.AddNew<br>rs1("srollno") = 1<br>rs1("sname") = "Hardik Molia"<br>rs1("saddress") = "University Road"<br>rs1("scity") = "Rajkot"<br>rs1("spercentage") = 80<br>rs1.Update | rs1.AddNew<br>rs1.Fields(0) = 1<br>rs1.Fields(1) = "Hardik Molia"<br>rs1.Fields(2) = "University Road"<br>rs1.Fields(3) = "Rajkot"<br>rs1.Fields(4) = 80<br>rs1.Update |

3rd method:-

```
Dim varfields, varvalues
varfields=Array("srollno","sname","saddress","scity","spercentage")
varvalues=Array(1,"Hardik Molia","University Road","Rajkot",80)
rs.AddNew varfields,varvalues
```

- **Delete** method deletes the current record or a group of records.

rs1.Delete

RecordSet supports tow ways to update the database
Immediate updating – updation of a single record to the data base
Batch updating – updation of multiple records to the data base

- **Update** method saves all changes made to a single record in a RecordSet. This method will not work if the RecordSet does not support updates.

rs1.Update

- **UpdateBatch** method is used to save all changes in a RecordSet to the database. This method is used when you are working on a RecordSet in batch update mode. If we want to modify 100 records, then instead of executing Update method for every record, it is better to execute UpdateBatch method at the end of the updation of 100 records. RecordSet's LockType must be set to adLockBatchOptimistic to use this method.

rs1.UpdateBatch

- **Save** method saves a RecordSet object to a file or a Stream object. When the save method is finished, the record pointer will point at the first record of the RecordSet.

RecordSet.Save destination,persistformat

rs1.save "a.rst"                  c:/windows/system32/a.rst
rs1.save "c:\a.rst",1             c:/a.rst xml format

Default format is Advanced Data TableGram Format(0)
XML Format(1)
rs1.open "c:\a.rst"
It is also possible to read data from the saved file.

| Parameter | Description |
|---|---|
| destination | Optional. Specifies where to save the RecordSet object path of the file. |
| persistformat | specifies the format of the RecordSet (XML or ADTG). |

- **Close** method closes the RecordSet object and releases the memory associated with it.

## 7.5 COMMAND OBJECT

The Command object is also used to execute SQL statements or open tables. But the main purpose of using Command object is to execute the stored procedures.

set objCommand=Server.CreateObject("ADODB.command")

### 7.5.1 Properties of Command Object:-

- **ActiveConnection** property associates an Active Connection object with the Command object. The Command object uses the Connection object assigned to this property.

- **CommandText** property sets or returns a string that contains a SQL statement, a table name, or a call to a stored procedure.

Set comm=Server.CreateObject("ADODB.Command")
comm.CommandText="select * from students"

- **CommandType** property specifes which kind of execution we want to perform using a Command object.

| Constant | Value | Description |
| --- | --- | --- |
| adCmdText | 1 | A SQL statement or a call to stored procedure. |
| adCmdTable | 2 | A table name. ADO uses internally generated SQL query. |
| adCmdStoredProc | 4 | A stored procedure. |
| adCmdUnknown | 8 | Default. |

- **CommandTimeout** property sets or returns the number of seconds, ASP engine waits while executing a command, after this time ASP engine cancels the execution attempt and generates an error. Default is 30 seconds.

### 7.5.2 Methods of Command Object:-

- **Execute** method executes a SQL statement or a stored procedure specified in the CommandText property of the Command object. The results are stored in a new RecordSet object.

167

```
objcommand.Execute
Example:-
Set comm = Server.CreateObject("ADODB.Command")

Set comm.ActiveConnection = con1

comm.CommandType = 1

comm.CommandText = "select * from students"

Set rs1 = comm.Execute
```

## 7.6 CURSOR TYPES

In ADO, a cursor is a set of rows. When we execute a select query, database returns rows of data. The resulting data is handled using a cursor. A cursor can be maintained either on the client or on the server.

**CursorType** property specifies how we can navigate records in RecordSet. It also specifies whether is it possible to see the database changes made by other users or not.

In ADO there are 4 different cursor types defined:

| Constant | Value | Description |
|---|---|---|
| adOpenForwardOnly | 0 | Default. We can move the record pointer in forward direction only. |
| adOpenKeyset | 1 | All types of record pointer movement are possible.<br><br>Only changes and deletions by other users are visible. Additions by other users are not visible. |
| adOpenDynamic | 2 | All types of record pointer movement are possible.<br><br>Additions, changes, and deletions by other users are visible |
| adOpenStatic | 3 | All types of record pointer movement are possible.<br><br>Additions, changes, and deletions by other users are not visible. |

### adOpenForwardOnly

The forward only cursor is the fastest cursor type. It does not suppor the RecordCount property and MovePrevious method of RecordSet. This cursor is mostly used with adLockReadOnly to read the data.

### adOpenKeyset

The key set cursor supports navigation of records in both the directions(forware and backward). It also allows us to insert, delete and update of records of a RecordSet and all changes made by other users will be accessible. But the new records inserted by other users will not be accessible. The cursor also supports bookmarking of records.

### adOpenDynamic

The dynamic cursor supports of navigation of records in both the directions(forware and backward). It also allows us to insert, delete and update of records of a RecordSet and all changes including insertions made by other users will be accessible. The cursor also supports bookmarking of records.

### adOpenStatic

The static cursor is the only cursor which is avallable when we maintain curson on client side. The server sends the data to the client. After this, there will be no communication from the server to the client. It also allows us to insert, delete and update of records of a RecordSet .The client can send the changes back to the server. The data is stored on client's memory so it reduces the load on server. The cursor supports navigation of records in both the directions(forware and backward).

If other client changes the same data, other clients will receive no notification of the changes from the server. So it may produce data inconsistency if there are many users updating the data simultaneously.

## 7.7 LOCK TYPES

Cursor location and cursor type specify how our data is going to be handled. The lock type specifies how we are going to protect the database from inconsistency.

**LockType** property specifies the type of locking scheme while editing a record of a RecordSet. This property is required to set properly to avoid data inconsistency in database.

| Constant | Value | Description |
|---|---|---|
| adLockReadOnly | 1 | Read-only records. Editing is not possible. |
| adLockPessimistic | 2 | locks a record immediately after editing |
| adLockOptimistic | 3 | locks records only when calling update method |

| adLockBatchOptimistic | 4 | Locks all records until the execution of BatchUpdate completes. |
|---|---|---|

### adLockReadOnly

This is the default lock type. This lock type is sued when we want to read the data only to display or for calculations. There is no possiblity to insert, delete or update the data.

### adLockPessimistic

When multiple users are modifying the same data and the possiblity of concurrent access is very high, we use this lock. The rows will be locked(inaccessible to other users) as soon as we begin making changes and will not be unlocked until the update method is called. adLockPessimistic is not supported if CursorLocation is set to adUseClient.

### adLockOptimistic

When the possiblity of concurrent access is low, we use this lock. With optimistic lock, the table or row will be locked only when the Update method of RecordSet is called. this will ensure successfully changes but will not prevent other users from changing the same data while we are modifying it. There will be no lock in the duration in which we update the values and call the Save or Update method.

### adLockBatchOptimistic

This lock caches all the changes locally until we execute the RecordSet's UpdateBatch method. This lock is useful when we want to do changes in batch mode. In Batch update all changes(insert, delete, update) will be pushed to the server in a group. Similar to the optimistic lock, There will be no lock in the duration in which we update the values and call the UpdateBatch method.

## 7.8 COMMUNICATING WITH DATABASE USING ODBC

Open Database Connectivity (ODBC) is a software interface to access the database. ODBC supports DSN based or DSN less connectivity.

### 7.8.1 Database Source Name

A Data Source Name (DSN) is a data structure that contains the database connection information. ODBC uses a DSN to connect to the database. A DSN may reside either in a registry or in a separate text file. It stores information such as:

Name of the data source
Directory of the data source
Name of a driver which can access the data source
User ID for database access (if required)
User password for database access (if required)

Three kinds of DSN exist:

User DSN (sometimes called machine DSN)

System DSN

File DSN

**User DSN:-**User DSN is available to use only for the user who created it. If some other user logs onto the computer, the user DSN will be unavailable. The computer stores User DSNs in the current user section of the Registry.

**System DSN:-**System DSN is available to use for all the users on a computer. No matter who created it and who wants to use it. The computer stores system DSNs in the local machine section of the Registry.

**File DSN:-**The computer stores file DSNs in a file(*.dsn). We can use these files on local computer or can use on any other computer by copying them.

**Steps to create a DSN:-**

1- open control panel
2- open administrative tools
3- open Data Sources (ODBC)

This will display the following applet.

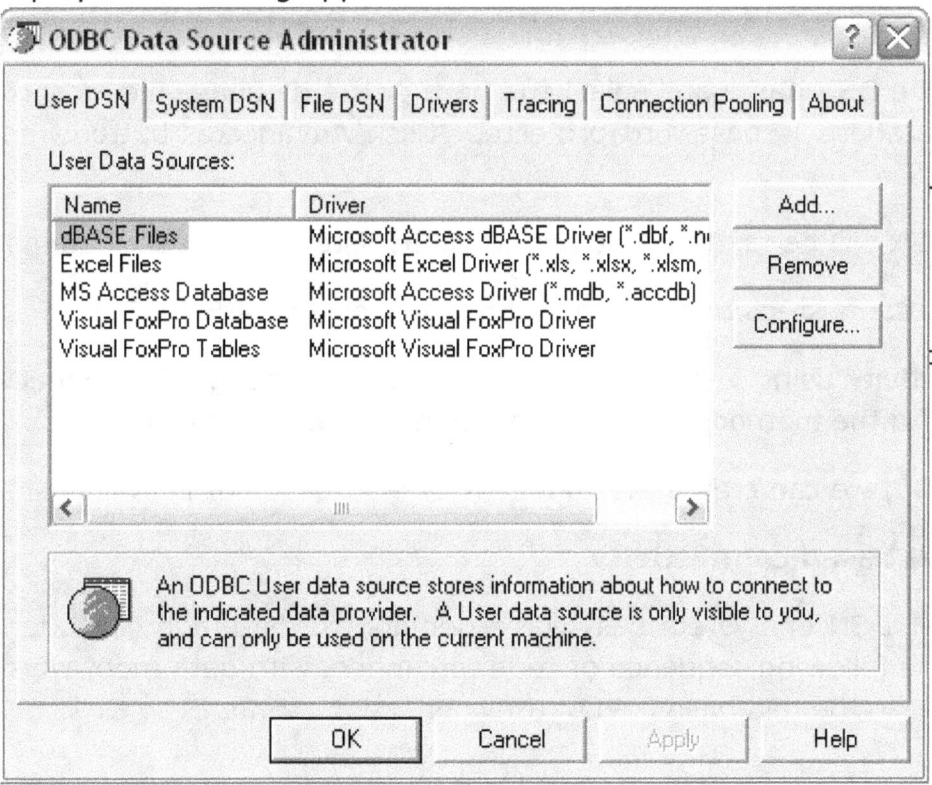

Select any of the User DSN, System DSN or File DSN according to the type of DSN you want to create. Let's create a User DSN.

4- For user data source list, select the database for Access files select "MS Access Database". If your required database is not listed in it, click on the Add button and select the database.

5- Click on Configure button to add a new User DSN.

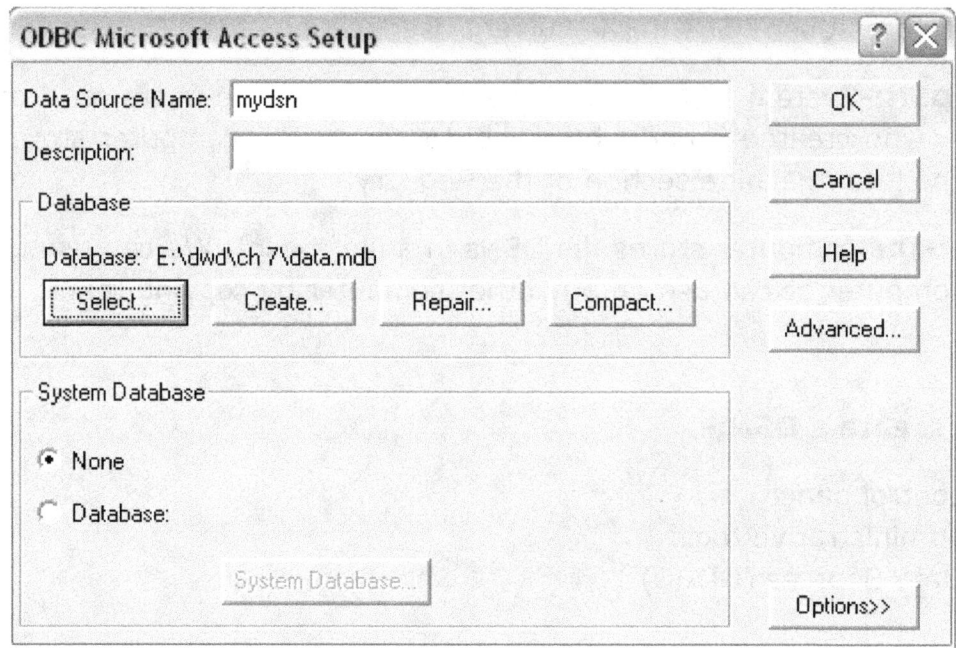

6- Enter the DSN name and select the path of the database file as shown above. If your database is password protected. Click Advanced.. button and enter it. Press ok.

7- Your DSN will be ready to use and it will be listed in the user data sources.

Follow the same steps to create a system DSN or a file DSN.

To connectivity using various methods we need to change the connection string according to the method. All other statements remain unchanged.

Using ODBC, we can create the connectivity DSN based or DSN less.

### 7.8.2 DSN based connectivity.

If the User DSN or System DSN corresponding to data.mdb is studentdata, we can use the following sequence of code to connect with data.mdb. we specify the table name in the database field of the connection string.

```
<%
Set con1 = Server.CreateObject("ADODB.Connection")
Con1.Open "dsn=studentdata;database=students"
Set rs1 = con.Execute("select * from students")
...........................
%>
```

if the file DSN corresponding to data.mdb is filedsnstudents, we can use the following sequence of code to connect with data.mdb

```
<%
Set con1 = Server.CreateObject("ADODB.Connection")
Con1.Open "filedsn=filedsnstudents;database=students"
Set rs1 = con.Execute("select * from students")
...........................
%>
```

### 7.8.3 DSN less connectivity

Instead of specifying the connection information through a DSN, we can also specify directly to the connection string as shown below.

```
<%
Set con1 = Server.CreateObject("ADODB.Connection")
con.Open                     "DRIVER={Microsoft                  Access
river(*.mdb)};DBQ=E:\database\data.mdb"
Set rs1 = con.execute("select * from students")
...........................
%>
```

## 7.9 COMMUNICATING WITH DATABASE USING OLEDB

Microsoft has designed OLE DB (Object Linking and Embedding, Database) interface to access various database uniformly. It is the replacement of ODBC. While ODBC supports only relational databases, OLEDB supports relational and non relational databases. We can specify the connection information either through a .udl (universal data link) file or directly in the code.

### 7.9.1 Using .UDL File.

Steps to create a .udl file

1 - Create a new blank .udl file from notepad. Select All files as the file type.

2 - Double click it it will open a dialog to enter connection information.

173

3 – From the provider tab, select Microsoft Jet 4.0 OLEDB Provider for Access. You can select any other if you want to connect with other database. Press next

3 – From the connection tab, Browse your .mdb file(Access file) and if it is password protected, enter it.

4 - Test connection.

Suppose we have given the name of the .udl file to db1, we can connect with the database as shown below.

db1.udl corresponds to data.mdb

Set con1 = Server.CreateObject("ADODB.Connection")

con.Open "file name=E\database\db1.udl"

Set rs1 = con.execute("select * from students")

.............................

### 7.9.2 Using Connection String.

It is also possible to connect with the database without using .udl file. We can specify the connection information directly into the connection string as shown below.

Set con1 = Server.CreateObject("ADODB.Connection")

con.open "Provider=Microsoft.Jet.OLEDB.4.0;Data Source=E:\.data.mdb"

Set rs1 = con.execute("select * from students")

.............................

Note:- WE WILL USE THE FOLLOWING DATA FOR STUDENTS TABLE IN DATA.MDB

| srollno | sname | saddress | scity | spercentage |
|---|---|---|---|---|
| 1 | RAJESH MODI | UNIVERSITY RC | RAJKOT | 85 |
| 2 | AJAY PATEL | KALAWAD ROA | RAJKOT | 85 |
| 3 | NIRAV PANDYA | JAGNATH PLOT | RAJKOT | 56 |
| 4 | PRADEEP BHATT | COLLEGE ROAD | MORBI | 22 |
| 5 | RAVI HIRPARA | PANCHESHWAI | JAMNAGAR | 89 |
| 6 | SUNIL MAHETA | MOTI BAG | JUNAGADH | 24 |
| 7 | ANIL SHAH | SANKUL ROAD | AMRELI | 64 |
| 8 | HARSH TRIVEDI | HOSPITAL ROA | JAMNAGAR | 86 |
| 9 | ALPESH HIRPARA | COLLEGE ROAD | AMRELI | 78 |
| 10 | NILESH PANDYA | GAUTAM ROAL | MORBI | 33 |

## 7.10 READING DATABASE

To read data from the database, we follow following steps.

1. Establish the connection with the database either using ODBC or OLEDB.

2. Execute a SELECT statement. we have three options(Connection object's Execute method, RecordSet object's Open method or Command object's Execute method)

3. Navigate the returned RecordSet object and display the data using Response.Write

### (READING DATABASE USING RECORDSET'S OPEN METHOD)

7.3.asp (DISPLAY ALL STUDENTS)

```
<HTML>
<BODY>
<%
Set con1 = Server.CreateObject("ADODB.Connection")
Set rs1 = Server.CreateObject("ADODB.RecordSet")

con1.open "Provider=Microsoft.Jet.OLEDB.4.0;Data Source=" &
Server.MapPath(".") & "\data.mdb"
q  = "select * from students"

rs1.Open q,con1

Response.Write "<H3>Student Information<H3>"
Response.Write "<TABLE BORDER=1>"
Response.Write "<TR><TH>ROLL NO<TH>NAME<TH>
ADDRESS<TH> CITY<TH>RESULT</TR>"

While Not rs1.EOF

Response.write "<TR>"
Response.Write "<TD>" & rs1.Fields(0)
Response.write "<TD>" & rs1.Fields(1)
Response.Write "<TD>" & rs1.Fields(2)
Response.Write "<TD>" & rs1.Fields(3)
Response.Write "<TD>" & rs1.Fields(4)
rs1.MoveNext

WEnd

rs1.Close
con1.Close
%>
</BODY></HTML>
```

**OUTPUT:-**

**Student Information**

| ROLL NO | NAME | ADDRESS | CITY | RESULT |
|---------|------|---------|------|--------|
| 1 | RAJESH MODI | UNIVERSITY ROAD | RAJKOT | 85 |
| 2 | AJAY PATEL | KALAWAD ROAD | RAJKOT | 85 |
| 3 | NIRAV PANDYA | JAGNATH PLOTS | RAJKOT | 56 |
| 4 | PRADEEP BHATT | COLLEGE ROAD | MORBI | 22 |
| 5 | RAVI HIRPARA | PANCHESHWAR TOWER | JAMNAGAR | 89 |
| 6 | SUNIL MAHETA | MOTI BAG | JUNAGADH | 24 |
| 7 | ANIL SHAH | SANKUL ROAD | AMRELI | 64 |
| 8 | HARSH TRIVEDI | HOSPITAL ROAD | JAMNAGAR | 86 |
| 9 | ALPESH HIRPARA | COLLEGE ROAD | AMRELI | 78 |
| 10 | NILESH PANDYA | GAUTAM ROAD | MORBI | 33 |

Description:-

- con1 is the Connection object. rs1 is the RecordSet object. q is the string to store SQL statement.
- We have used Server object's MapPath method to get the absolute path to the current directory in which data.mdb is stored.
- After opening a rs1, we will use a while loop with the termination condition is Not rs1.EOF
- Initially record pointer points to the first record and so rs1.EOF is false. With each iteration, while loop prints all the fields using rs1's Fields collection and moves the rs1 to the next record using its MoveNext method.
- When record pointer will be at last record and rs1.MoveNext will try to move to next record, it will set the EOF property to True and so that will be the end of the loop.
- It is also possible to access the fields directly by using their field names. We can replace the loop body with the following. It will generate the same result.

```
While Not rs1.EOF
Response.write "<TR>"
Response.Write "<TD>" & rs1("SROLLNO")
Response.write "<TD>" & rs1.Fields("SNAME")
Response.Write "<TD>" & rs1.Fields("SADDRESS")
Response.Write "<TD>" & rs1.Fields("SCITY")
Response.Write "<TD>" & rs1.Fields("SPERCENTAGE")
rs1.MoveNext
WEnd
```

## (READING DATABASE USING CONNECTION'S EXECUTE METHOD)
7.4.asp (DISPLAY ROLL NO AND PERCENTAGE OF ONLY PASS STUDENTS)

```
<HTML>
<BODY>
<%
Set con1 = Server.CreateObject("ADODB.Connection")
con1.open        "Provider=Microsoft.Jet.OLEDB.4.0;Data
Source=" _
& Server.MapPath(".") & "\data.mdb"

q  = "select * from students where spercentage > 35"

Set rs1 = con1.Execute(q)
Response.Write "<H3>Passed Students<H3>"
Response.Write "<TABLE BORDER=1>"
Response.Write                          "<TR><TH>ROLL
NO<TH>RESULT</TR>"

While Not rs1.EOF

Response.write "<TR>"
Response.Write "<TD>" & rs1("SROLLNO")
Response.Write "<TD>" & rs1("SPERCENTAGE")
rs1.MoveNext

WEnd

rs1.Close
con1.Close
%>
</BODY>
</HTML>
```

**OUTPUT:-**

**Passed Students**

| ROLL NO | RESULT |
|---------|--------|
| 1 | 85 |
| 2 | 85 |
| 3 | 56 |
| 5 | 89 |
| 7 | 64 |
| 8 | 86 |
| 9 | 78 |

Note:-

We have used rs1("SROLLNO") and rs1("SPERCENTAGE"). We can also access the values using Fields collection as we read in example 7.3. sometimes we can get the same result by using different SQL statements. We want to get the rollno and percentage of only passed students. Here are the two SQL statements to get this data

1:- select * from students where spercentage > 35

when we will execute this statement, we will get all the information of those students whose percentages are more than 35.

Because of the "*", if we want to use Fields collection, we need to follow the order of the columns as per the database table.

So we can replace

rs1("SROLLNO") by rs1.Fields(0) and rs1("SPERCENTAGE") by rs1.Fields(4)

2:- select srollno, spercentage from students where spercentage > 35

if we use this statement, we will get only two columns in the RecordSet which are srollno and spercentage. And so in this case, we can replace

rs1("SROLLNO") by rs1.Fields(0) and rs1("SPERCENTAGE") by rs1.Fields(1)

So the conclusion is that, the RecordSet's Fields collection follows the indexing of the columns retrieved as a result of a SQL query. And so in case of selected columns retrieval, the indexing of columns in RecordSet is different than the indexing of columns in actual table.

## (READING DATABASE USING COMMAND'S EXECUTE METHOD)

### 7.4.asp (DISPLAY ROLL NO AND PERCENTAGE OF ONLY FAIL STUDENTS)

```
<HTML><BODY>
<%
Set con1 = Server.CreateObject("ADODB.Connection")
Set com1 = Server.CreateObject("ADODB.Command")

con1.open          "Provider=Microsoft.Jet.OLEDB.4.0;Data
Source=" _
& Server.MapPath(".") & "\data.mdb"

q   = "select srollno,spercentage from students where
spercentage<= 35"

Set com1.ActiveConnection = con1
com1.CommandType = 1
com1.CommandText = q
Set rs1 = com1.Execute

Response.Write "<H3>Failed Students<H3>"
Response.Write "<TABLE BORDER=1>"
Response.Write "<TR><TH>ROLL NO<TH>RESULT</TR>"

While Not rs1.EOF
Response.write "<TR>"
Response.Write "<TD>" & rs1.Fields(0)
Response.Write "<TD>" & rs1.Fields(1)
rs1.MoveNext
WEnd

rs1.Close
con1.Close%>
</BODY></HTML>
```

OUTPUT:-

**Failed Students**

| ROLL NO | RESULT |
|---------|--------|
| 4       | 22     |
| 6       | 24     |
| 10      | 33     |

Note:-

We can get the same results by using

1 Connection's Execute method or RecordSet's Open method or Command's Execute method.

2 Different SQL statements.

3 Access RecordSet's fields using its Fields collection or directly by field name.


It is also possible to get data from more than two tables. We just need to write a proper SQL statement and access statements from the columns returned by that statement.


## 7.11 WRITING DATABASE

To write data (insert new record) to the database, we follow following steps.

1. Establish the connection with the database either using ODBC or OLEDB.

2. Execute a INSERT statement or use RecordSet's AddNew method.

3. View the record in the database.

4. If we use, RecordSet's AddNew method, we must have to open the RecordSet using the cursor mode and lock mode which support insertion.

### (ADD A RECORD USING INSERT STATEMENT WITH COMMAND OBJECT)

7.6.asp (ADD A STUDENT RECORD)

```
<HTML>
<BODY>
<%
Set con1 = Server.CreateObject("ADODB.Connection")
Set com1 = Server.CreateObject("ADODB.Command")

con1.open "Provider=Microsoft.Jet.OLEDB.4.0;Data Source=" _
& Server.MapPath(".") & "\data.mdb"

q = "insert into students values(11,'AJAY VORA','NAVRANGPURA','AHMEDABAD',54)"

Set com1.ActiveConnection = con1
com1.CommandType = 1
com1.CommandText = q
Set rs1 = com1.Execute
%>
</BODY>
</HTML>
```

**A new record of roll no 11 will be added to the students table.**
**Note: - If we have read values of a record from user. We must have to write the SQL statement with proper syntax.(string values inside single quotations)**
**The same program can be written as follow with variables.**

```
<HTML>
<BODY>
<%
Set con1 = Server.CreateObject("ADODB.Connection")
Set com1 = Server.CreateObject("ADODB.Command")

con1.open "Provider=Microsoft.Jet.OLEDB.4.0;Data Source=" _
& Server.MapPath(".") & "\data.mdb"

sturoll=11
stuname="AJAY VORA"
stuaddr="NAVRANGPURA"
stucity="AHMEDABAD"
stuper=54

q = "insert into students values(" & sturoll & ",'" & stuname & "','" & stuaddr & "','" &
_
stucity & "'," & stuper & ")"

Set com1.ActiveConnection = con1
com1.CommandType = 1
com1.CommandText = q
Set rs1 = com1.Execute
%>
</BODY>
</HTML>
```

**Note:-**
After putting variable values and concatenation,
"insert into students values(" & sturoll & ",'" & stuname & "','" & stuaddr & "','" & _
stucity & "'," & stuper & ")"

becomes,

"insert into students values(11,'AJAY VORA','NAVRANGPURA','AHMEDABAD',54)"

## (ADD A RECORD USING ADD NEW METHOD OF RECORDSET OBJECT)

7.7.asp (ADD A STUDENT RECORD)

```
<HTML>
<BODY>
<%
Set con1 = Server.CreateObject("ADODB.Connection")
Set rs1 = Server.CreateObject("ADODB.RecordSet")

con1.open "Provider=Microsoft.Jet.OLEDB.4.0;Data Source=" & Server.MapPath(".")
& "\data.mdb"

rs1.Open "students",con1,2,2
rs1.AddNew
rs1("srollno")=11
rs1("sname")="AJAY VORA"
rs1("saddress")="NAVRANGPURA"
rs1("scity")="AHMEDABAD"
rs1("spercentage")=54
rs1.Update
rs1.Close
con1.Close

%>
</BODY>
</HTML>
```

**A new record of roll no 11 will be added to the students table.**


**Note:-**

**Here also we can access the columns using Fields collection of RecordSet object.**

**rs1.Open "students",con1,2,2 will open the table students with adOpenDynamic(2) cursor and adLockPessimistic(2) lock. So we can update the database. We can use other combination of cursor type and lock type also.**

## 7.12 UPDATING DATABASE

To update data (modify existing record) to the database, we follow following steps.

1. Establish the connection with the database either using ODBC or OLEDB.

2. Execute a UPDATE statement or use RecordSet's Update or BatchUpdate method.

3. View the record in the database.

4. If we use, RecordSet's Update or UpdateReocrd method, we must have to open the RecordSet using the cursor mode and lock mode which support editing.

## (UPDATE A RECORD USING UPDATE STATEMENT WITH COMMAND OBJECT)

7.8.asp (UPDATE PERCENTAGE OF AJAY VORA TO 76)

```
<HTML>
<BODY>
<%
Set con1 = Server.CreateObject("ADODB.Connection")
Set com1 = Server.CreateObject("ADODB.Command")

con1.open "Provider=Microsoft.Jet.OLEDB.4.0;Data Source=" _
& Server.MapPath(".") & "\data.mdb"

q = "update students set spercentage=76 where sname='AJAY VORA'"

Set com1.ActiveConnection = con1
com1.CommandType = 1
com1.CommandText = q
Set rs1 = com1.Execute

%>
</BODY>
</HTML>
```
**The percentage of AJAY VORA will be updated**

## (UPDATE A RECORD USING UPDATE METHOD OF RECORDSET OBJECT)

7.9.asp (UPDATE PERCENTAGE OF AJAY VORA TO 76)

```
<HTML>
<BODY>
<%
Set con1 = Server.CreateObject("ADODB.Connection")
Set rs1 = Server.CreateObject("ADODB.RecordSet")

con1.open "Provider=Microsoft.Jet.OLEDB.4.0;Data Source=" & Server.MapPath(".")
& "\data.mdb"

rs1.Open "select * from students where sname='AJAY VORA'",con1,2,2

rs1("spercentage")=76
rs1.Update

rs1.Close
con1.Close

%>
</BODY>
</HTML>
```

**Note:-**
**For update also, we need to open the RecordSet with proper lock type and cursor type. As discussed in previous example.**
**It is also possible to fill the RecordSet with all the students. Move to the record which we want to update and update it. This is shown in below example. It  updates the percentage of AJAY VORA to 76**

```
<HTML>
<BODY>
<%
Set con1 = Server.CreateObject("ADODB.Connection")
con1.open "Provider=Microsoft.Jet.OLEDB.4.0;Data Source=" & Server.MapPath(".")
& "\data.mdb"

set rs1 = server.CreateObject("ADODB.RecordSet")
rs1.Open "select * from students",con1,2,2

While Not rs1.EOF

        If rs1("sname")="AJAY VORA" Then
                rs1("spercentage")=76
                rs1.Update
        End If
        rs1.MoveNext
WEnd
rs1.Close
con1.Close%></BODY></HTML>
```

## 7.13 DELETING DATABASE

To delete data (delete existing record) to the database, we follow following steps.

1. Establish the connection with the database either using ODBC or OLEDB.

2. Execute a DELETE statement or use RecordSet's Delete method.

3. View the database.

4. If we use, RecordSet's Delete method, we must have to open the RecordSet using the cursor mode and lock mode which support deleting.

## (DELETE A RECORD USING DELETE STATEMENT WITH COMMAND OBJECT)

7.10.asp (DELETE RECORD OF ROLL NO 11)

```
<HTML>
<BODY>
<%
Set con1 = Server.CreateObject("ADODB.Connection")
Set com1 = Server.CreateObject("ADODB.Command")
con1.open "Provider=Microsoft.Jet.OLEDB.4.0;Data Source=" &
Server.MapPath(".") & "\data.mdb"
q = "delete from students where srollno=11"
Set com1.ActiveConnection = con1
com1.CommandType = 1
com1.CommandText = q
Set rs1 = com1.Execute
%>
</BODY>
</HTML>
```
**The percentage of AJAY VORA will be updated**

## (DELETE A RECORD USING DELETE METHOD OF RECORDSET OBJECT)

7.11.asp (DELETE RECORD OF ROLL NO 11)

```
<HTML>
<BODY>
<%
Set con1 = Server.CreateObject("ADODB.Connection")
Set rs1 = Server.CreateObject("ADODB.RecordSet")

con1.open "Provider=Microsoft.Jet.OLEDB.4.0;Data Source=" &
Server.MapPath(".") & "\data.mdb"
```

184

```
rs1.Open "select * from students where srollno=3",con1,2,2

rs1.Delete
rs1.Close
con1.Close
%>
</BODY>
</HTML>
```

**Note:-**

**For delete also, we need to open the RecordSet with proper lock type and cursor type. As discussed in previous example.**

**It is also possible to fill the RecordSet with all the students. Move to the record which we want to delete and delete it. This is shown in below example. It deletes the record of roll no 11.**

```
<HTML>
<BODY>
<%
Set con1 = Server.CreateObject("ADODB.Connection")
Set rs1 = Server.CreateObject("ADODB.RecordSet")

con1.open "Provider=Microsoft.Jet.OLEDB.4.0;Data Source=" &
Server.MapPath(".") & "\data.mdb"


rs1.Open "select * from students",con1,2,2

While Not rs1.EOF


        If rs1("srollno")=11 Then
                rs1.Delete
        End If
        rs1.MoveNext
WEnd

rs1.Close
con1.Close

%>
</BODY>
</HTML>
```

## 7.14 EXAMPLES

## (WRITE A PROGRAM TO DISPLAY THE RESULT OF A STUDENT.) (BASED ON THE ROLL NO ENTERED BY HIM.)

7.12.asp (ENTER ROLL NUMBER)

```
<HTML>
<BODY>
<CENTER>
<FORM ACTION=7.12result.asp METHOD=POST>
<H3>ENTER ROLL NO:-</H3>
<INPUT TYPE=TEXT SIZE=3 NAME=ROLL><BR>
<INPUT TYPE=SUBMIT VALUE=RESULT>
</FORM>
</BODY>
</HTML>
```

**OUTPUT:-**

**ENTER ROLL NO:-**

```
3
RESULT
```

7.12result.asp (DISPLAY RESULT)

```
<HTML>
<BODY>
<%
Set con1 = Server.CreateObject("ADODB.Connection")
Set rs1 = Server.CreateObject("ADODB.RecordSet")

con1.open "Provider=Microsoft.Jet.OLEDB.4.0;Data Source=" &
Server.MapPath(".") & "\data.mdb"

q  = "select srollno,sname,spercentage from students where srollno=" &
Request.Form("ROLL")

rs1.Open q,con1,3,3

If rs1.RecordCount=1 Then

Response.Write "<H3>Student Result<H3>"
Response.Write "<TABLE BORDER=1>"
Response.Write "<TR><TH>ROLL NO<TH>NAME<TH>RESULT</TR>"
Response.write "<TR>"
Response.Write "<TD>" & rs1("srollno")
Response.write "<TD>" & rs1("sname")
```

```
Response.Write "<TD>" & rs1("spercentage")


Else
Response.Write "Invalid RollNo"
Response.Write "<BR><INPUT TYPE=BUTTON onClick=history.back()
VALUE=BACK>"
End If

rs1.Close
con1.Close
%>
</BODY>
</HTML>
```

OUTPUT:-

## Student Result

| ROLL NO | NAME | RESULT |
|---------|------|--------|
| 3 | NIRAV PANDYA | 56 |

**Description:-**

A student opens 7.12.asp file containing the form with a textbox for entering roll no and submit button. We have set the ACTION attribute to 7.12result.asp so when a student will submit the form for the result, 7.12result.asp will be executed.

7.12result.asp will get the roll no entered by the student in Form collection of the Request object because of the Post method of form submission.

So we have used Request.Form("ROLL") where ROLL is the name of the textbox in which student enters his roll no, in the generation of query statement.

We can also assign Request.Form("ROLL") to a separate variable, and we can use that variable in query.

Finally RecordSet will get only one record because of the where condition in query and 7.12result.asp will print the result in a tabular form.

If student has entered an invalid roll no, database returns nothing and so RecordCount property of RecordSet's object will be 0. so we checked RecordCount property. If it is 1 then it prints the result otherwise it gives an error message and a button to back to the entry form.

## (WRITE A PROGRAM TO INSERT A NEW STUDENT RECORD.) (BASED ON A FILLED FORM.)

7.13.asp (ENTER STUDENT DETAILS)

```
<HTML>
<BODY>
<CENTER>
<FORM ACTION=7.13insert.asp METHOD=POST>
<H3>ENTER STUDENT DETAILS:-</H3>
<TABLE>
<TR><TD>ROLLNO:-<TD><INPUT TYPE=TEXT SIZE=3 NAME=STUROLL>
<TR><TD>NAME:-<TD><INPUT TYPE=TEXT NAME=STUNAME>
<TR><TD>ADDRESS:-<TD><INPUT TYPE=TEXT NAME=STUADDR>
<TR><TD>CITY:-<TD><INPUT TYPE=TEXT NAME=STUCITY>
<TR><TD>PERCENTAGE:-<TD><INPUT TYPE=TEXT NAME=STUPER>
<TR><TD><TD><INPUT TYPE=SUBMIT VALUE=ADD>
</TABLE>
</FORM>
</BODY>
</HTML>
```

OUTPUT:-

### ENTER STUDENT DETAILS:-

| | |
|---|---|
| ROLLNO:- | 7 |
| NAME:- | JAY PANDYA |
| ADDRESS:- | RANI TOWER |
| CITY:- | RAJKOT |
| PERCENTAGE:- | 65 |

ADD

7.13insert.asp (ADD A STUDENT RECORD)

```
<HTML>
<BODY>
<%
Set con1 = Server.CreateObject("ADODB.Connection")
Set rs1 = Server.CreateObject("ADODB.RecordSet")

con1.open "Provider=Microsoft.Jet.OLEDB.4.0;Data Source=" &
Server.MapPath(".") & "\data.mdb"

rs1.Open "students",con1,3,3

rs1.AddNew
```

```
rs1("srollno")=Cint(Request.Form("sturoll"))
rs1("sname")=Request.Form("stuname")
rs1("saddress")=Request.Form("stuaddr")
rs1("scity")=Request.Form("stucity")
rs1("spercentage")=Cint(Request.Form("stuper"))
rs1.Update

Response.Write "Record is inserted"
rs1.Close
con1.Close
%>
</BODY>
</HTML>
```

**OUTPUT:-**

Record is inserted

Note:- we discussed client side and server side validation in chapter 6. in real life applications, we should implement all the programs with validation and formatting. For simplicity of examples, I have not added code for validation and formatting.

Example 7.14

Data.mdb has one table users(uname,upwd) which stores pairs of valid usernames and passwords.

Modify the example 6.20 from chapter 6. Validate the username and password entered by the user from the database. If both are valid then only allow user to visit other content otherwise generate an error message.

### 7.14login.asp (ENTER USERNAME AND PASSWORD)

```
<HTML>
<BODY>
<FORM ACTION=7.14page1.ASP METHOD=POST>
<TABLE>

<TR><TD>USERNAME:-
<TD><INPUT TYPE="TEXT" NAME="UNAME" SIZE=10>
</INPUT><BR>

<TR><TD>PASSWORD:-
<TD><INPUT TYPE="PASSWORD" NAME="PWD" SIZE=10>
</INPUT>

<TR><TD><INPUT TYPE = "SUBMIT" VALUE="Login">
<TD><INPUT TYPE = "RESET">
</TABLE>
</FORM></BODY></HTML>
```

**OUTPUT:-**

USERNAME:- hardikmolia

PASSWORD:- *******

[ Login ]        [ Reset ]

## 7.14page1.asp (VALIDATION PAGE)

```
<HTML>
<BODY>
<%
Dim user,password
user = Request.Form("UNAME")
password = Request.Form("PWD")
Set con1 = Server.CreateObject("ADODB.Connection")
Set rs1 = Server.CreateObject("ADODB.RecordSet")

con1.open "Provider=Microsoft.Jet.OLEDB.4.0;Data Source=" &
Server.MapPath(".") & "\data.mdb"
q  = "select * from users where uname='" & user & "' and upwd='" & password &
"'"
rs1.Open q,con1,3,3

If rs1.RecordCount=1 Then
Session("Username") = user
Session("logintime") = Now
Response.Write "<A HREF=7.14page2.asp>Logout</A>"
Else
Response.Write "Invalid username or password."
Response.Write "<INPUT TYPE=BUTTON VALUE=BACK onClick=history.back()>"
End If

rs1.Close
con1.Close
%>
</FORM>
</BODY>
</HTML>
```

**OUTPUT:- If username and password are valid**

Logout

**OUTPUT:- If username or/and password is/are invalid**

Invalid username or password. [ BACK ]

## 7.14page2.asp (LOGOUT PAGE)

```
<HTML>

<BODY>

<%

Response.Write "Good Bye " & Session("username")

Response.Write "<BR>Your logged in period:- " &
DateDiff("n",Session("logintime"),Now)

_& " Minutes."

%>

</FORM>

</BODY>

</HTML>
```

**OUTPUT:-**

Good Bye hardikmolia
Your logged in period:- 2 Minutes.

## Example 7.15

**Data.mdb has one table emps(empid,salary). Give 10% increment to all employees using UpdateBatch method.**

## 7.15.asp (UpdateBatch Example)

```
<HTML>
<BODY>
<%
Set con1 = Server.CreateObject("ADODB.Connection")
Set rs1 = Server.CreateObject("ADODB.RecordSet")
con1.open        "Provider=Microsoft.Jet.OLEDB.4.0;Data        Source="    &
Server.MapPath(".") & "\data.mdb"
rs1.Open "select * from emps",con1,3,3

While Not rs1.EOF
rs1("salary") = rs1("salary") + ((rs1("salary")*10)/100)
rs1.MoveNext
Wend
rs1.UpdateBatch
```

```
Response.Write "DATABASE IS UPDATED"
rs1.Close
con1.Close
%>
</BODY>
</HTML>
```

**OUTPUT:-**

DATABASE IS UPDATED

Note:- With Update method, we can write rs1.Update inside the loop. As we are execute rs1.MoveNext it will update the data automatically so we can also avoid use of Update and UpdateBatch.

## Example 7.16

**Store the student information in XML format using RecordSet's Save method.**

**7.16save.asp (Save RecordSet to XML file)**

```
<HTML>
<BODY>
<%
Set con1 = Server.CreateObject("ADODB.Connection")
Set rs1 = Server.CreateObject("ADODB.RecordSet")

con1.open "Provider=Microsoft.Jet.OLEDB.4.0;Data Source=" &
Server.MapPath(".") & "\data.mdb"

rs1.Open "select * from students",con1
rs1.Save "c:\myrs.xml",1

rs1.Close
con1.Close

%>
```
**Check** c:\myrs.xml for database

**7.16open.asp (Open RecordSet saved as XML file without Connection object)**

```
<HTML>
<BODY>
<%
set rs1 = server.CreateObject("ADODB.RecordSet")
rs1.Open "c:\myrs.xml"
Response.Write "<H3>Student Information<H3>"
Response.Write "<TABLE BORDER=1>"
```

```
Response.Write                                    "<TR><TH>ROLL
NO<TH>NAME<tH>ADDRESS<TH>CITY<TH>RESULT</TR>"

While Not rs1.EOF
Response.write "<TR>"
Response.Write "<TD>" & rs1.Fields(0)
Response.write "<TD>" & rs1.Fields(1)
Response.Write "<TD>" & rs1.Fields(2)
Response.Write "<TD>" & rs1.Fields(3)
Response.Write "<TD>" & rs1.Fields(4)
rs1.MoveNext
WEnd

rs1.Close
%>
</BODY>
</HTML>
```

OUTPUT:-

### Student Information

| ROLL NO | NAME | ADDRESS | CITY | RESULT |
|---------|------|---------|------|--------|
| 1 | RAJESH MODI | UNIVERSITY ROAD | RAJKOT | 85 |
| 2 | AJAY PATEL | KALAWAD ROAD | RAJKOT | 85 |
| 3 | NIRAV PANDYA | JAGNATH PLOTS | RAJKOT | 56 |
| 4 | PRADEEP BHATT | COLLEGE ROAD | MORBI | 22 |
| 5 | RAVI HIRPARA | PANCHESHWAR TOWER | JAMNAGAR | 89 |
| 6 | SUNIL MAHETA | MOTI BAG | JUNAGADH | 24 |
| 7 | ANIL SHAH | SANKUL ROAD | AMRELI | 64 |
| 8 | HARSH TRIVEDI | HOSPITAL ROAD | JAMNAGAR | 86 |
| 9 | ALPESH HIRPARA | COLLEGE ROAD | AMRELI | 78 |
| 10 | NILESH PANDYA | GAUTAM ROAD | MORBI | 33 |

Example 7.17

Design a page which provides following facilities

1- Search a student record based on his roll no and display his information in a table.

2 – Delete a student based on his roll no.

3 – Update a student record based on his roll no. to update data, show the old values in text boxes.

193

## 7.17.asp (Search, Delete and Modify)

```
<HTML>
<BODY>
<FORM ACTION=7.17.ASP METHOD=POST>
<CENTER>
STUDENT ROLLNO.<INPUT TYPE=TEXT SIZE=3 NAME=ID></INPUT>
SELECT OPTION:-
<SELECT NAME="OPTION">
<OPTION>SEARCH</OPTION>
<OPTION>DELETE</OPTION>
<OPTION>MODIFY</OPTION>
</SELECT>

<INPUT TYPE="SUBMIT" VALUE="GO">
</FORM>

<%
Set cat = Request.Form("OPTION")
Set id = Request.Form("ID")
Set con1=Server.CreateObject("ADODB.Connection")
Set rs1 =Server.CreateObject("ADODB.RecordSet")
con1.Open "Provider=Microsoft.Jet.OLEDB.4.0;Data Source=" &
server.mappath(".") & "\data.mdb"
q = "Select * from students where srollno = " & Cint(id)
rs1.Open q, con1, 3, 3
If rs1.BOF Then
Response.Write "<BR><BR>No student found for rollno " & id
Response.End
End If

If cat="SEARCH" Then
Response.Write "<TABLE BORDER=1>"
Response.Write "<TR><TD>ROLLNO:-<TD>" & rs1("SROLLNO")
Response.Write "<TR><TD>NAME:-<TD>" & rs1("SNAME")
Response.Write "<TR><TD>ADDRESS:-<TD>" & rs1("SADDRESS")
Response.Write "<TR><TD>CITY:-<TD>" & rs1("SCITY")
Response.Write "<TR><TD>PERCENTAGE:-<TD>" & rs1("SPERCENTAGE")
End If

If cat="DELETE" Then
rs1.Delete
Response.Write "<BR><BR>Data of RollNo:- " & id & " is deleted..."
End If

If cat="MODIFY" Then %>
<BR><BR>
<FORM NAME=F1 METHOD="POST" ACTION="7.17update.ASP"><TABLE
BORDER=0>
```

```
<BR><TR><TD>ROLLNO<TD><INPUT TYPE=TEXT NAME=sturoll
VALUE=<%=rs1("SROLLNO")%>>

<BR><TR><TD>NAME<TD><INPUT TYPE=TEXT NAME=stuname
VALUE=<%=rs1("SNAME")%>>

<BR><TR><TD>ADDRESS<TD><INPUT TYPE=TEXT NAME=stuaddress
VALUE=<%=rs1("SADDRESS")%>>

<BR><TR><TD>CITY<TD><INPUT TYPE=TEXT NAME=stucity
VALUE=<%=rs1("SCITY")%>>

<BR><TR><TD>PERCENTAGE<TD><INPUT TYPE=TEXT NAME=stupercentage
VALUE=<%=rs1("SPERCENTAGE")%>>

<INPUT TYPE=SUBMIT VALUE="UPDATE">
<%
Response.Cookies("sID") = id
End If
%>
</BODY>
</HTML>
```

**OUTPUT AFTER SEARCH**

STUDENT ID. [    ]   SELECT OPTION:- [SEARCH ▼]  [ GO ]

| | |
|---|---|
| ROLLNO:- | 1 |
| NAME:- | RAJESH MO |
| ADDRESS:- | UNIVERSITY ROAD |
| CITY:- | RAJKOT |
| PERCENTAGE:- | 85 |

SEARCH
DELETE
MODIFY

**OUTPUT AFTER DELETION**

STUDENT ID. [    ]   SELECT OPTION:- [SEARCH ▼]  [ GO ]

Data of RollNo:- 1 is deleted...

**OUTPUT DURING MODIFICATION**

195

```
STUDENT ID. [    ]  SELECT OPTION:- [SEARCH ▾] [ GO ]


          ROLLNO      [2                    ]
          NAME        [NISHANT SHARMA       ]
          ADDRESS     [ISCON                ]
          CITY        [AHMEDABAD            ]
          PERCENTAGE  [67]              [ UPDATE ]
```

## 7.17update.asp (Update Data)

```
<HTML>
<BODY>
<%
Set con1=Server.CreateObject("ADODB.Connection")
Set rs1=Server.CreateObject("ADODB.Recordset")

con1.open "Provider=Microsoft.Jet.OLEDB.4.0;Data Source=" &
server.mappath(".") & "\data.mdb"

q= "Select * from students where SROLLNO = " &  Request.Form("sturoll")

rs1.open q, con1, 3, 3

rs1("SROLLNO") = Request.Form("sturoll")
rs1("SNAME") = Request.Form("stuname")
rs1("SADDRESS") = Request.Form("stuaddress")
rs1("SCITY") = Request.Form("stucity")
rs1("SPERCENTAGE") = Request.Form("stupercentage")
rs1.Update

Response.Write "Record is updated."
%>

</BODY>
</HTML>
```

**OUTPUT:-**

Record is updated.

Description:-

The program is the mixture of all the important things which we have studied so far. It is easy to understand the basic flow of the above program. It is better to run the program once and then it will be easy to understand how the program works.

It is also possible to include the code of modification in 7.17.asp. We just need to decide whether is it a new request or a update request at the beginning of 7.17.asp we can use any method of state management (cookies or session variables)

## 7.15 REVIEW QUESTIONS

1. Explain AdRotator component.

2. Explain BrowserCapabilities component.

3. Explain ADO objects.

4. Explain Connection object's open and execute methods.

5. Explain RecordSet object's BOF, EOF, bookmark, CursorType, Locktype and RecordCount properties.

6. Explain RecordSet object's Open method.

7. How to do navigation of records using a recordset?

8. Explain RecordSet object's AddNew, Delete, Update, UpdateBatch and Save methods.

9. Explain Command object's ActiveConnection, CommandType and CommandText properties.

10. Explain Command object's Execute method.

11. Explain various cursor types.

12. Explain various lock types.

13. Explain types of DSNs.

14. Write steps to do DSN based database connectivity using ODBC.

15. Write steps to do DSN less database connectivity using ODBC.

16. Write steps to do database connectivity using OLEDB with connection string.

16. Write steps to do database connectivity using OLEDB with UDL file.

www.ingramcontent.com/pod-product-compliance
Lightning Source LLC
Chambersburg PA
CBHW081046170526
45158CB00006B/1869